Hydrogen-Based Energy Conversion

Hydrogen-Based Energy Conversion

Polymer Electrolyte Fuel Cells and Electrolysis

Editor

Jin-Soo Park

MDPI • Basel • Beijing • Wuhan • Barcelona • Belgrade • Manchester • Tokyo • Cluj • Tianjin

Editor
Jin-Soo Park
Sangmyung University
Korea

Editorial Office
MDPI
St. Alban-Anlage 66
4052 Basel, Switzerland

This is a reprint of articles from the Special Issue published online in the open access journal *Energies* (ISSN 1996-1073) (available at: https://www.mdpi.com/journal/energies/special_issues/ Hydrogen_Based_Energy_Conversion).

For citation purposes, cite each article independently as indicated on the article page online and as indicated below:

LastName, A.A.; LastName, B.B.; LastName, C.C. Article Title. *Journal Name* **Year**, *Volume Number*, Page Range.

ISBN 978-3-0365-0690-6 (Hbk)
ISBN 978-3-0365-0691-3 (PDF)

© 2021 by the authors. Articles in this book are Open Access and distributed under the Creative Commons Attribution (CC BY) license, which allows users to download, copy and build upon published articles, as long as the author and publisher are properly credited, which ensures maximum dissemination and a wider impact of our publications.

The book as a whole is distributed by MDPI under the terms and conditions of the Creative Commons license CC BY-NC-ND.

Contents

About the Editor vii

Preface to "Hydrogen-Based Energy Conversion" ix

Seohee Lim and Jin-Soo Park
Composite Membranes Using Hydrophilized Porous Substrates for Hydrogen Based
Energy Conversion
Reprinted from: *Energies* **2020**, *13*, 6101, doi:10.3390/en13226101 1

Jong-Hyeok Park and Jin-Soo Park
KOH-doped Porous Polybenzimidazole Membranes for Solid Alkaline Fuel Cells
Reprinted from: *Energies* **2020**, *13*, 525, doi:10.3390/en13030525 15

Do-Hyeong Kim and Moon-Sung Kang
Pore-Filled Anion-Exchange Membranes with Double Cross-Linking Structure for Fuel Cells
and Redox Flow Batteries
Reprinted from: *Energies* **2020**, *13*, 4761, doi:10.3390/en13184761 27

**Riccardo Balzarotti, Saverio Latorrata, Marco Mariani, Paola Gallo Stampino and
Giovanni Dotelli**
Optimization of Perfluoropolyether-Based Gas Diffusion Media Preparation for PEM Fuel Cells
Reprinted from: *Energies* **2020**, *13*, 1831, doi:10.3390/en13071831 41

Reza Omrani and Bahman Shabani
Gas Diffusion Layers in Fuel Cells and Electrolysers: A Novel Semi-Empirical Model to Predict
Electrical Conductivity of Sintered Metal Fibres
Reprinted from: *Energies* **2019**, *12*, 855, doi:10.3390/en12050855 55

Guo-Bin Jung, Shih-Hung Chan, Chun-Ju Lai, Chia-Chen Yeh and Jyun-Wei Yu
Innovative Membrane Electrode Assembly (MEA) Fabrication for Proton Exchange Membrane
Water Electrolysis
Reprinted from: *Energies* **2019**, *12*, 4218, doi:10.3390/en12214218 73

Chan-Ho Song and Jin-Soo Park
Effect of Dispersion Solvents in Catalyst Inks on the Performance and Durability of Catalyst
Layers in Proton Exchange Membrane Fuel Cells
Reprinted from: *Energies* **2019**, *12*, 549, doi:10.3390/en12030549 83

Xuyang Zhang, Andrew Higier, Xu Zhang and Hongtan Liu
Experimental Studies of Effect of Land Width in PEM Fuel Cells with Serpentine Flow Field and
Carbon Cloth
Reprinted from: *Energies* **2019**, *12*, 471, doi:10.3390/en12030471 93

Ashley Fly, Kyoungyoun Kim, John Gordon, Daniel Butcher and Rui Chen
Liquid Water Transport in Porous Metal Foam Flow-Field Fuel Cells: A Two-Phase Numerical
Modelling and Ex-Situ Experimental Study
Reprinted from: *Energies* **2019**, *12*, 1186, doi:10.3390/en12071186 103

About the Editor

Jin-Soo Park received a B.Sc. in Environmental Engineering from the Inha University, Incheon, Republic of Korea (1998), an M.Sc. in Environmental Science and Engineering from the Gwangju Institute of Science and Technology, Gwangju, Republic of Korea (2000), and his Ph.D. in Environmental Science and Engineering from the Gwangju Institute of Science and Technology, Gwangju, Republic of Korea (2004). He is presently a Full Professor of Green Chemical Engineering at Sangmyung University, Cheonan, Republic of Korea, where he currently leads a research group involved in electromembrane processes and electrochemical engineering, such as fuel cells, water electrolysis, capacitive deionization, electrodialysis/reverse electrodialysis, electrochemical oxidation processes, and redox flow batteries, and has served as head of the Future Environment and Energy Research Institute, Sangmyung University.

Preface to "Hydrogen-Based Energy Conversion"

Hydrogen-based energy conversion from chemical to electrical energy, and vice versa, is one of the most promising energy paradigms. Two technologies, fuel cells and electrolysis, play a crucial role in solving the emission of greenhouse gas and other pollutants from the combustion of hydrocarbon fuels. Hydrogen could be ultimately used as carbon-free fuel, which can be produced by water electrolysis powered by renewable energy such as wind, solar, ocean, and so on; can be stored by compression, liquefaction, adsorption, or chemical conversion of hydrogen; can be distributed by pipelines, tank trailers, and so on; and, finally, can be utilized by fuel cells. Both of the hydrogen-based technologies seek to decrease internal resistance in order to obtain a performance that is as high as possible. Ion-conducting polymers are a great material that can be used make thinner and more durable electrolytes so as to produce efficient stacks and systems with good specific and volumetric power density. There are two different types of ion-conducting polymers, cation (mainly proton) and anion exchangeable polymers. The former leads to anodic and cathodic reactions in acidic conditions, but the latter is in a basic condition. The difference decides the type of electrocatalysts. In an acidic condition, platinum is mainly used as anodic and cathodic electrocatalysts. Some non-platinum electrocatalysts could be used in a basic condition. Furthermore, the difference also causes different types of electrodes and a different environment to affect mass transport of gas, ions (liquid), and electrons in gas diffusion layers and/or the flow field of bipolar plates. Thus, numerical simulation and precise characterization techniques are of significant importance for predicting and analyzing the difference.

This highly interdisciplinary book containing many aspects of science and engineering in polymer electrolyte membrane fuel cells includes nine valuable papers. The tremendous efforts are greatly appreciated. I wish that this book helps all readers understand better this rapidly developing and potentially helpful technology.

Jin-Soo Park
Editor

Article

Composite Membranes Using Hydrophilized Porous Substrates for Hydrogen Based Energy Conversion

Seohee Lim and Jin-Soo Park *

Department of Green Chemical Engineering, College of Engineering, Sangmyung University, Cheonan 31066, Korea; qook0258@gmail.com
* Correspondence: energy@smu.ac.kr; Tel.: +82-41-550-5315

Received: 20 October 2020; Accepted: 19 November 2020; Published: 21 November 2020

Abstract: Poly(tetrafluoroethylene) (PTFE) porous substrate-reinforced composite membranes for energy conversion technologies are prepared and characterized. In particular, we develop a new hydrophilic treatment method by in-situ biomimetic silicification for PTFE substrates having high porosity (60–80%) since it is difficult to impregnate ionomer into strongly hydrophobic PTFE porous substrates for the preparation of composite membranes. The thinner substrate having ~5 μm treated by the gallic acid/(3-trimethoxysilylpropyl)diethylenetriamine solution with the incubation time of 30 min shows the best hydrophilic treatment result in terms of contact angle. In addition, the composite membranes using the porous substrates show the highest proton conductivity and the lowest water uptake and swelling ratio. Membrane-electrode assemblies (MEAs) using the composite membranes (thinner and lower proton conductivity) and Nafion 212 (thicker and higher proton conductivity), which have similar areal resistance, are compared in I–V polarization curves. The I–V polarization curves of two MEAs in activation and Ohmic region are very identical. However, higher mass transport limitation is observed for Nafion 212 since the composite membrane with less thickness than Nafion 212 would result in higher back diffusion of water and mitigate cathode flooding.

Keywords: composite membrane; perfluorinated sulfonic acid; ionomer; electrolyte; fuel cell

1. Introduction

The unstable crude oil prices and global warming caused by greenhouse gases drive us to use alternative energy sources. Companies and governments have made substantial investments in new and renewable energy over the past few years [1]. Among the new and renewable energy, hydrogen had begun to be translated into the alternative energy source area due to easy deployment into electricity infrastructure, diversified energy sources to produce hydrogen from fossil fuels to biomass, improvement of local air pollution, and recent matured technologies such as fuel cells and electrolyzers [2,3]. Using the electrochemistry-driven energy conversion technologies, hydrogen could be produced from water by using electricity and could be consumed to convert into the water along with generating electricity ($H_2 + O_2 \leftrightarrow H_2O$) [4–6].

Fuel cells are the most promising technology utilizing hydrogen. Proton exchange membrane fuel cells (PEMFCs) using hydrogen as fuel show high efficiency during the direct conversion of chemical energy to electric energy. In addition, there is no greenhouse gas emission when hydrogen is produced by water electrolysis using electricity supplied from renewable energy sources such as wind, solar, biomass, etc. [1,7]. Nevertheless, the installation cost is still higher than conventional energy conversion technology such as internal combustion engine and the technical level of durability and reliability must be raised to a higher level to enter the full-fledged fuel cell market [8–10]. Membrane-electrodeassembly (MEA) is a key component in PEMFCs, which consists of a piece of proton exchange membrane (PEM) sandwiched between two catalyst layers as an electrode [11].

The recent approach for PEM development has shifted from more proton conductivity to more proton conductance (in other words, from less resistivity to less areal specific resistance) at low relative humidity and higher temperature. Thus, the development of thinner PEMs is crucial to minimize areal specific resistance by the trade-off between less thickness and mechanical/chemical stability. It directly results in a significant decrease in Ohmic losses which is an increase in stack power density and a decrease in material cost [10]. Mechanical reinforcement of thin PEMs (ca. 10–20 μm) is one of the approaches to overcome less mechanical/chemical stability by thinning the thickness. The reinforcement could be achieved by the development of porous substrate-reinforced composite membranes. It could be done only by filling high-conductivity ionomers into porous substrates or by forming a three-layered structure (ionomer/ionomer filled substrate/ionomer) [12–20]. Perfluorosuflonic acid (PFSA) ionomers are still the most frequently used material even though less expensive hydrocarbon membranes have been intensively developed [7,21–23] Among many reasons for the use of PFSA ionomers, the main one would be the good stability against mechanical and chemical stress occurring during fuel cell or water electrolysis operation. Since the less volume of PFSA is used compared to non-reinforced PFSA membranes, benefits can be obtained in terms of material cost [12,24,25]. To prepare reinforced composite membranes, PFSA ionomer dispersion normally in a mixed solvent of water and alcohols and poly(tetrafluoroethylene) (PTFE) porous substrate is often used due to high proton conductivity of the ionomers and polymeric compatibility with PFSA, respectively [26–28]. Nevertheless, the preparation process of porous PTFE-reinforced PFSA composite membranes is very difficult since not all the hydrophilic ionomer dispersions are compatible with hydrophobic PTFE porous substrates. The fabrication of the composite membranes with an incomplete filling of PFSA ionomers into the substrates causes them to lose mechanical strength and chemical stability as well as gas permeability [29]. However, few studies to discuss the effect of materials for hydrophilic treatment of porous substrates for the preparation of porous PTFE-reinforced PFSA composite membranes have been reported.

The hydrophilization of the hydrophobic material surface could be attained by oxygen plasma, UV radiation, grafting, surface oxidation by strong acids, hydrolysis, coating, or lamination. Hydrophilization treatment with plasma or UV radiation is effective because it directly exposes energy to the hydrophobic material surface. However, the hydrophilization using plasma has a limitation in that the process must be performed in a vacuum state, and the one using long-term or strong UV radiation may damage the material surface to irreversibly change the properties of materials. In addition, there is a limit to completely hydrophilize the inner pores of porous substrates. Similarly, lamination is not good for materials with complicated structure. The coating is lower than the plasma and UV radiation in terms of durability, but it is much simpler than the aforementioned methods and inexpensive. Thus, it is frequently used in the industrial hydrophilization process [30]. Initially, the increase in hydrophilicity of PTFE surfaces was obtained by surfactants, but there are too many parameters to be considered for good wettability [31,32]. Recently, biomimetic materials have been deposited on porous hydrophobic microfiltration/ultrafiltration membranes for better water flux from the oil-in-water emulsion and protein wastewater [33]. It is found that the pyrogallol moiety in gallic acid (GA) with amino-terminated substances (ATS) such as siloxane generated a similar mussel-inspired adhesive coating via Michael addition/Schiff base reactions in alkaline conditions [34–36]

Herein, an approach to increase the wettability of hydrophobic PTFE substrates is investigated by using the nature-born materials, i.e., mussel-inspired silicified polysiloxane adhesive materials, to overcome the aforementioned incompatibility between hydrophobic PTFE substrates and hydrophilic ionomer dispersions. Hydrophilization on porous PTFE substrates (porosity 40-90%) from the polymerization of GA with respect to the ATS, i.e., 3-aminopropyltriethoxysilane (APTES), N-[3-(trimethoxysilyl)propyl]ethylenediamine (TMPEDA), and (3-trimethoxysilylpropyl)diethylenetriamine (TMPDETA) is carried out, not on microfiltration (MF) or ultrafiltration (UF) which is less porous (porosity <40%) than PTFE. The properties of composite membranes using the porous substrates hydrophilically treated by GA and one of the amino-terminated substances are discussed in terms of water contact angle of hydrophilically treated porous substrates and proton conductivity of composite membranes. Afterward, fuel cell performance of composite membranes is measured and discussed to investigate the effect of the biomimetic coating materials on the properties of porous PTFE-reinforced composite membranes.

2. Materials and Methods

2.1. Materials

Aeos™ ePTFE and PTU0214210 PTFE substrates were purchased from Zeus, the USA, and Sterlitech, USA, respectively. The main physical properties are summarized in Table 1. The 3,4,5-trihydroxybenzoic acid (GA), APTES, TMPEDA, and TMPDETA were obtained from Sigma Aldrich. Their chemical structures are shown in Table 2. For pH 8.5 buffer, trizma® hydrochloride (Tris-HCl) was purchased from Dongin Biotech Co., Ltd., South Korea. All the chemicals were used as received without further purification.

Table 1. Specifications of poly(tetrafluoroethylene) (PTFE) substrates used in this study.

Substrates	Zeus Aeos™ ePTFE	Sterlitech PTU0459010
thickness (μm)	~5	~25
porosity (%)	~80	~60
pore size (μm)	0.2-0.5	0.45

Table 2. Chemical structures of 3,4,5-trihydroxybenzoic acid (GA), 3-aminopropyltriethoxysilane (APTES), N-[3-(trimethoxysilyl)propyl]ethylenediamine (TMPEDA), and (3-trimethoxysilylpropyl)diethylenetriamine (TMPDETA).

Chemicals	GA	ATS		
		APTES	TMPEDA	TMPDETA
chemical structure	(structure)	(structure)	(structure)	(structure)

2.2. Hydrophilic Coating of GA/ATS on PTFE Substrates

PTFE substrates were immersed in acetone and ethanol for 6 h, respectively, and then dried in the air prior to hydrophilic treatment. GA of 0.2 g was dissolved in 10 mM Tris-HCl of 100 mL for a GA solution, and three different ATS solutions of 0.13 M using APTES (coded as #1), TMPEDA (coded as #2), and TMPDETA (coded as #3) were prepared in ethanol for ATS solutions. GA/ATS solutions were prepared by mixing 100 mL of the GA solution and 20 mL of the corresponding ATS solution. For the coating of hydrophilic materials, PTFE substrates were immersed in zipper bags filled by the corresponding solutions placed on a plate orbital shaker and were incubated at room temperature for a certain time period under shaking, followed by rinsing with distilled water and drying in the air. The abbreviated incubation conditions are summarized in Table 3.

Table 3. Incubation conditions of PTFE substrates and their abbreviations.

Samples	Coating Materials	GA/ATS (mL/mL)	Incubation Time (min)
#1_60	GA/APTES	5/1	60
#1_120	GA/APTES	5/1	120
#1_360	GA/APTES	5/1	360
#2_60	GA/TMPEDA	5/1	60
#2_120	GA/ TMPEDA	5/1	120
#2_360	GA/ TMPEDA	5/1	360
#3_60	GA/TMPDETA	5/1	60
#3_120	GA/TMPDETA	5/1	120
#3_360	GA/TMPDETA	5/1	360

2.3. Preparation of Porous PTFE-Reinforced Composite Membranes

Reinforced composite membranes were prepared by hydrophilized PTFE substrates and D1021 Nafion™ dispersion (EW 1100, 10 wt.%, Chemours). As shown in Figure 1, the hydrophilically treated porous substrate provides better wettability for the Nafion™ dispersion. However, the untreated pristine one completely repels the dispersion of the substrate surface. The procedure consists of six steps as follows: (1) the PTFE substrates were immersed into the dispersion for 5 min; (2) the membranes were placed on a glass plate doctor-bladed by a film applicator with a thickness of 10 μm; (3) the membranes were dried in a convective oven at 70 °C for 10 min; (4) another 10 μm-layer Nafion was coated on another side of the membranes; (5) the membranes were dried again in a vacuum oven at 70 °C for 6 h; (6) finally the membranes were annealed in a convective oven at 190 °C for 12 min. The scheme of the preparation is illustrated in Figure 2. All the composite membranes were boiled in a 0.5 M H_2SO_4 solution for 6 h for protonation and boiled in distilled water for 6 h for removal of residual acid prior to use.

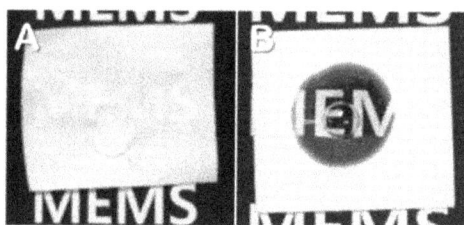

Figure 1. Wettability test of the Nafion™ dispersion on pristine (**A**) and GA/#3_60-treated PTFE porous substrates (**B**).

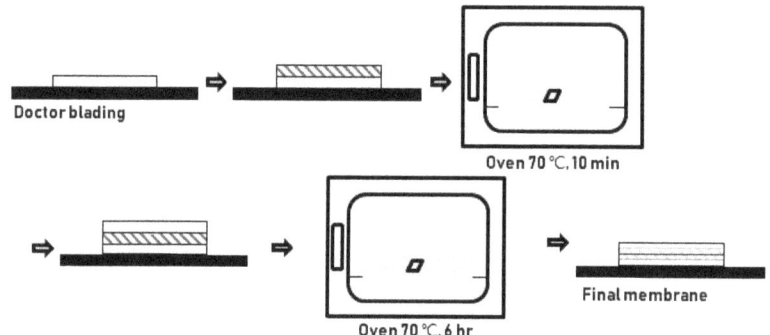

Figure 2. Schematic diagram of the preparation of porous PTFE-reinforced composite membranes.

2.4. Characterization of Material Structure and GA/ATS Solutions

Measurement of contact angle of PTFE substrates with and without hydrophilic material coating was carried out using a contact angle meter (Attension® Theta Lite, Bioline Scientific, Sweden) as soon as a droplet of the Nafion dispersion falls on substrates. Microcosmic morphology of the substrates and the composite membranes were obtained using a field emission scanning electron microscopy (FE-SEM) system (ZEISS Sigma 500, Germany). Attenuated total reflectance-Fourier transformed infrared spectroscopy (ATR-FTIR) characterization of samples was obtained by a JASCO FT-IR 4700 spectrometer (USA) with an ATR accessory containing a Ge crystal with a wavenumber resolution of 4 cm^{-1} and range of 600-4000 cm^{-1}. The microscopic state of GA/ATS solutions was detected by dynamic light scattering (DLS) (ELSZ-1000, Otsuka Electronics Co., Ltd., Japan).

2.5. Characterization of Porous PTFE-Reinforced Composite Membranes

Water uptake of composite membranes was measured by immersing the membranes into distilled water at room temperature for 12 h after the dry membranes (W_{dry}) were weighed. Afterward, the membranes were taken out, surface water was wiped out, and the wet weight (W_{wet}) was then measured. Finally, the water uptake of composite membranes was calculated by using the following equation [37]:

$$\text{Water uptake}(\%) = \frac{W_{wet} - W_{dry}}{W_{dry}} \times 100. \quad (1)$$

The swelling ratio is calculated by the following equation:

$$\text{Swelling ratio}(\%) = \frac{L_{wet} - L_{dry}}{L_{dry}} \times 100 \quad (2)$$

where L_{dry} and L_{wet} are the dimension (length, width, and thickness) of composite membranes.

The ion exchange capacity (IEC) of composite membranes is firstly soaked in 1.0 M H_2SO_4 for 24 h to replace functional groups with a proton. The excess acid solution on the surface of the membrane was thoroughly washed off with distilled water, and the membranes were immersed in 1.0 M NaCl (precisely 20 mL) for 24 h. The amount of proton ion-exchanged with Na^+ was measured by a titration method with 0.01 N NaOH by an auto-titrator (848 Titrino plus, Metrohm, Switzerland). The IEC was calculated by the following equation:

$$\text{IEC} = \frac{n \times M \times V}{m} \quad (3)$$

where n is the number of electrons gain or lost, M is the molar concentration of the titration solution (mmol/mL), V is the consumed volume of the titration solution (mL) and m is the weight of the dry membrane [38].

The transport number of composite membranes was measured by using the electromotive force (*emf*) method which is a method of estimating the transport number of conducting species in ionic and mixed conductors in a two-compartment cell with a composite membrane as a diaphragm in two separate solutions having different concentrations. First of all, composite membranes with a size of 2 × 2 cm^2 were immersed in 0.001 M NaCl for 24 h, were taken out, and were installed between two compartment cells which were filled with 0.001 and 0.005 M NaCl on both sides of the cells. Air bubbles on the membrane surfaces were completely removed. Finally, two Luggin capillaries with Ag/AgCl wires filled in a saturated KCl solution were mounted to the nearest locations to the installed membrane. Two wires exposed out of each Luggin capillary reference electrode were then connected to a digital voltage meter (34401A, Agilent, USA) to record voltage between two electrodes. The transport number of composite membranes was calculated by the following equation:

$$E_m = \frac{RT}{F}(1 - 2\overline{t_+}) \ln \frac{C_1}{C_2} \qquad (4)$$

where, E_m is the measured membrane potential, T is the temperature of the solution (K), F is the Faraday constant (96485 C eq^{-1}), R is the ideal gas constant (8.3145 J mol^{-1} K^{-1}), and C_1 is the lower electrolyte concentration, C_2 is the higher electrolyte concentration [38].

For the determination of proton conductivity of composite membranes, the electrical impedance of composite membranes was measured. The membranes were cut into a piece with 2×2 cm^2 and then were impregnated with 1 M H_2SO_4 for 12 h to replace functional groups into a proton. A piece of a sample was washed with distilled water and was placed in a four-electrode in-plane conductivity cell. The in-plane cell was immersed in distilled water, and the impedance value was measured at a frequency range from 10^6 to 10^{-3} Hz using a potentiostat/galvanostat with a frequency response analyzer (SP-150, BioLogic, France). Proton conductivity of composite membranes is calculated by the following equation:

$$\sigma \text{ (S/cm)} = \frac{L}{R \cdot A} \qquad (5)$$

where, L is the thickness of a membrane, R is the impedance of a membrane at zero phase angle and A is the cross-sectional area of a membrane.

2.6. Fuel Cell Performance Using Porous PTFE-Reinforced Composite Membranes

The membrane-electrode assemblies (MEAs) were made by a spraying technique using a commercially available Pt-based electrocatalyst (Alfa Aesar HiSPEC™ 4000, USA) with the catalyst loading of 0.4 mg cm^{-2} and Nafion® 212 (Chemours) or a lab-made composite membrane. MEAs prepared in this study were mounted in a unit cell with an electrode active area of 9 cm^2 and were evaluated in a test station (CNL Energy Co. Ltd., Republic of Korea). Hydrogen crossover was measured by using a test station (CNL Energy Co. Ltd., Republic of Korea) at 70 °C, with H_2/N_2 flows fixed at 0.2/0.2 L min^{-1} and 2.5 bar absolute pressure in the anode and cathode compartment. The cell operation conditions were in 100% relative humidity for both anode and cathode and the cell temperature of 70 °C. The fuel and oxidant were hydrogen and air to the anode and cathode with a stoichiometry ratio of 1.2 and 2.0, respectively.

3. Results and Discussion

3.1. Characterization of GA/ATS Solutions

Surface coating of adhesive hydrophilic materials is based on the synthesis of polysiloxane from GA/ATS, which is caused by adhesive deposition of the mussel-inspired coating [35]. To confirm in-situ biomimetic silicification, the reactions between GA with the catechol hydroxyl moiety and ATS with the amino moieties have been investigated. Firstly, the visual observation of GA/ATS solutions with incubation time was carried out. Figure 3 shows the co-deposition of GA and ATS. Color and transparency of the GA/#1 change and the co-deposit particles in the GA/#1 solution are not clearly formed. It is, however, observed that the GA/#2 and GA/#3 solutions successfully formed the co-deposition. Initially, both solutions are transparent and yellow. As the incubation time increases, the color of the solutions becomes opaque and dark brown by visual observation. This is also confirmed by the laser transmission of the synthetic solutions in Figure 3. In all cases, laser light penetrates clear solutions immediately after mixing GA and #2 but does not penetrate any more after 120, 30, and 30 min for the GA/#1, GA/#2, and GA/#3, respectively. In addition, the GA/#3 shows more scattered laser light penetration at the beginning compared to the GA/#1 and GA/#2. It infers that the homogeneous reaction is faster than the others. In the case of the GA/#3, after the incubation time of

120 min, laser light penetrates again since the synthetic material begins to grow massively and to sink, finally permitting laser light to transmit slightly again through the supernatant.

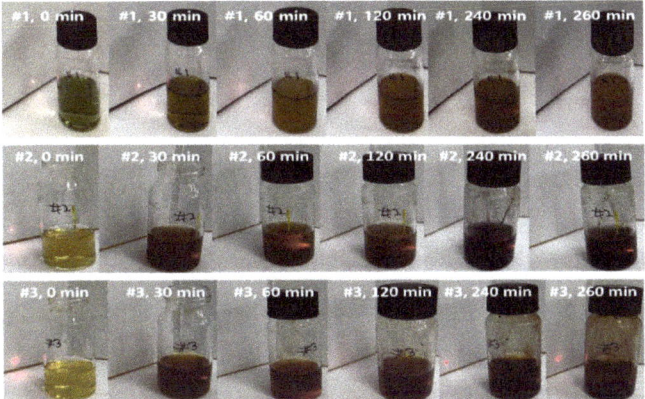

Figure 3. Visual observation of GA/ATS/alkaline buffer solutions with incubation time (#1): APTES; (#2): TMPEDA; (#3): TMPDETA.

The hydrophilic treatment of the porous PTFE-substrates could be attained by the self-coating of the adhesive polysiloxane on the surface of the two kinds of substrates when the substrates are immersed in the GA/ATS/alkaline buffer solutions. The self-coating results in a decrease in the contact angle of the surface of the substrates in a solid-water-air system. Normally, the surface is hydrophobic when the contact angle larger than 90° and hydrophilic when the contact angle is less than 90°. As shown in Figure 4, the treatment of GA and #1 results in no decrease in contact angle as the incubation time increases, which means that the surface of both substrates was not treated to be hydrophilic. It was reported that GA and #1 in alkaline buffer successfully deposited hydrophilic polysiloxane on MF and UF filters [33]. However, no significantly hydrophilic coating was attained for the substrates with higher porosity than MF and UF. In contrast, #2 and #3 with GA in alkaline buffer successfully achieved the hydrophilic coating on the substrates. Thus, the GA/#1 was excluded for further hydrophilic treatment. Figure 5 shows the change in the contact angle of the surface of two substrates treated by GA/#2 and GA/#3 with incubation time. Hydrophilicity in all the cases increased with the incubation time. The minimum incubation time for the increase was 30 min for both GA/ATS solutions. It was observed that there was no change in hydrophilicity prior to 30 min. A decrease rate in the contact angle for the thinner substrate (Zeus) was greater than the thicker substrate (Sterlitech). There is no distinct difference in hydrophilic coating for the thicker substrate with the type of the solutions, but better hydrophilic treatment by the GA/#3 for the thinner substrate is observed than that by the GA/#2. In addition, the contact angle of the GA/#3 treated thinner substrate only becomes less than 90° with an incubation time of 30 min. As a result, the combination of the thinner substrate having ~5 µm and the GA/#3 solution is effective for hydrophilic coating by in-situ biomimetic silicification. In particular, the ATS with the longer amino moieties results in better Michael addition and Schiff base reactions with the catechol hydroxyl moiety of GA. In addition, according to the results of the size distribution of co-deposits formed in the GA/#2 and GA/#3 solutions (see Figure 6 and Table 4), the GA/#3 forms narrower and smaller particles than the GA/#2 as the incubation time increases. Hence, the porous substrate immersed in the GA/#3 results in better hydrophilic treatment. It could be concluded that the hydrophilic treatment of the Zeus porous PTFE substrates in the GA/#3 solution for 30 min incubation time is optimal.

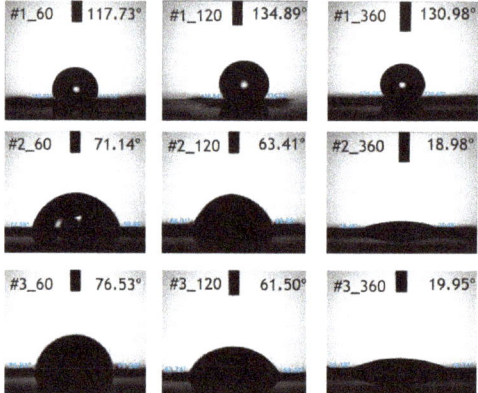

Figure 4. The contact angle of the surface of the porous PTFE substrate (Zeus) treated by GA and APTES (**#1**) at the first row, GA and TMPEDA (**#2**) at the second row, and GA and TMPDETA (**#3**) at the third row with the incubation time of 60, 120, and 360 min from the left-hand side, respectively.

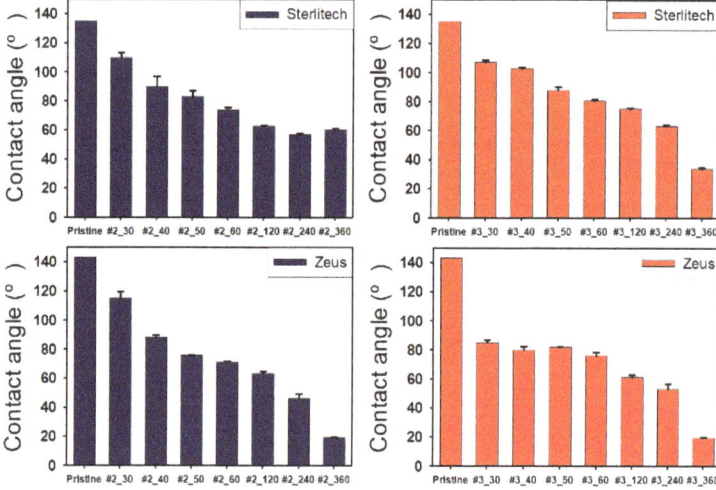

Figure 5. Change of contact angle of the surface of the porous PTFE substrate treated by GA and TMPEDA (**#2**) at the left-hand side (**blue**) and GA and TMPDETA (**#3**) at the right-hand side (**red**) with incubation time and the type of substrates (Sterlitech in the top row and Zeus in the bottom row).

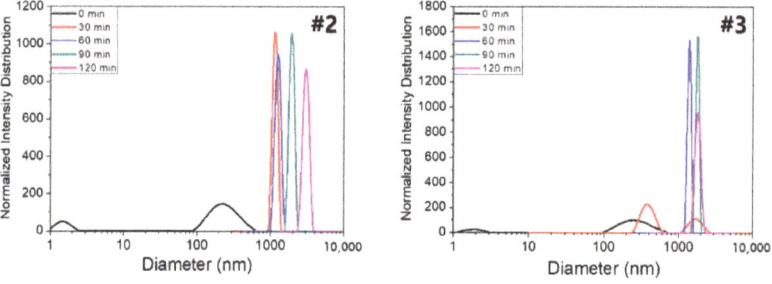

Figure 6. A dynamic light scattering of the GA/#2 (**left**) and GA/#3 (**right**) with incubation time.

Table 4. Average sizes of the particles formed in the GA/#2 and GA/#3 with incubation time.

Time (min)	Average Particle Sizes (nm)	
	GA/#2	GA/#3
0	185	205
30	1142	1050
60	1264	1377
90	1938	1781
120	3023	1810

As shown in Figure 7, the microscopic surface morphology of porous substrates shows an obvious coating layer after simple immersion in the GA/#2 and GA/#3 solutions in the corresponding SEM images. However, after the incubation time of 120 min, the coating material covers most of the surface of porous substrates. It could cause improper impregnation of PFSA ionomers into porous substrates due to the blocking of the surface pores of substrates by the covered hydrophilic material. As a result, Zeus porous substrates immersed in the GA/#3 show a significant decrease in contact angle less than 90° at the incubation time of 30 min due to the formation of smaller co-deposits.

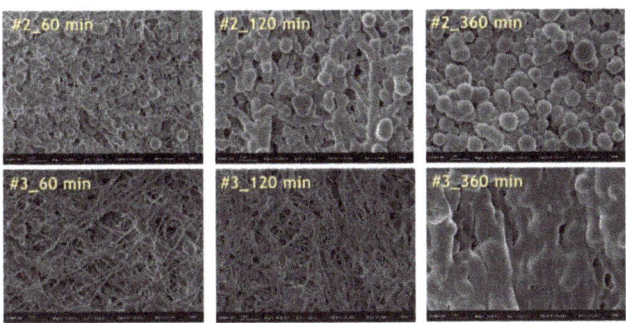

Figure 7. SEM images of the surface on the Zeus PTFE substrates after hydrophilic coating (GA/#2 in the top row and GA/#3 in the bottom row).

3.2. Characterization of GA/ATS-Treated Porous PTFE Substrate Based Composite Membranes

As determined earlier, the GA/#3 and Zeus porous substrates are only used for the preparation of composite membranes. Table 5 summarizes the results of the physical characterization of the composite membranes using GA/#3 treated Zeus porous substrates. All data in Table 5 are the average values of five samples with less than 10% standard deviation. The thickness of the composite membranes increased as the incubation time increased in all composite membranes. When the hydrophilically treated porous substrates were immersed in PFSA dispersion in the first step of the preparation of composite membranes, the porous substrates with higher incubation time absorbed more PFSA dispersion. It results in an increase in thickness for composite membranes with higher incubation time in the GA/#3. Thick composite membranes have thick skin layers on both sides of substrates since the thickness of substrates is similar. Accordingly, the water uptake of the composite membranes also increases as the incubation time increases. The main reason for using composite membranes in fuel cell applications is the suppression of swelling ratio in the area since the electrodes coated on polymeric electrolyte membranes could be cracked due to continuous areal expansion and contraction of membranes by hydration and dehydration during on and off cyclic operation of fuel cells [39]. Due to the skeleton effect of porous substrates, the variation of swelling ratio in the area is lower than that in thickness. Similarly, both swelling ratios also increase with the incubation time.

Table 5. Physical properties of composite membranes using Zeus porous substrates with incubation time in the GA/#3 solution.

Samples	Thickness (μm)	Water Uptake (%)	Swelling Ratio (%)	
			Area	Thickness
#3_30	27	11.6	21.87	12.50
#3_40	27	16.5	24.40	13.58
#3_50	30	17.7	25.48	16.67
#3_60	33	22.5	24.44	16.83
#3_120	38	25.6	24.38	20.69
#3_240	39	29.3	26.32	22.22
#3_360	40	35.9	29.74	23.53

Table 6 summarizes the results of the characterization of the composite membranes using GA/#3 treated Zeus porous substrates. All data in Table 6 are the average values of five samples with less than 6% standard deviation. As discussed earlier, the higher amount of hydrophilic co-deposits is coated on the substrate as the incubation time increases. Consequently, the inactive volume for proton transport increases, leading to a decrease in the ion-conducting fraction of the composite membranes which was comprised of pore-impregnated PFSA ionomer. Hence, it is observed in Table 6 that the ion conductivity of the composite membranes using the porous substrates with higher incubation time decreases. Similarly, the ion exchange capacity of the composite membranes also decreases with incubation time since the equivalent weight of the sulfonic acid group of PFSA ionomer in dry weight of composite membranes decreases. In addition, the transport number representing the ability of permselectivity of proton for the composite membranes using porous substrates treated by the GA/#3 for the incubation time greater than 120 min decreases substantially. It means that the composite membranes still have pores to allow to penetrate co-ions. It might be due to the coverage of hydrophilic co-deposits on the surface of the substrate or the filling of those into the substrate. The co-deposits could not provide permselectivity and result in a decrease in transport number.

Table 6. Electrochemical properties of composite membranes using Zeus porous substrates with incubation time in the GA/#3 solution.

Samples	Proton Conductivity (S cm^{-1})	Ion Exchange Capacity (meq g^{-1})	Transport Number (-)	Areal Resistance (Ω cm^2)
#3_30	0.069	0.843	0.99	0.039
#3_40	0.065	0.834	0.99	0.042
#3_50	0.064	0.826	0.99	0.047
#3_60	0.055	0.815	0.99	0.060
#3_120	0.053	0.789	0.98	0.072
#3_240	0.050	0.762	0.97	0.078
#3_360	0.049	0.742	0.85	0.082

Among the composite membranes, the GA/#3_30-treated composite membrane is chosen to compare with Nafion 212 as reference. As summarized in Table 7, the composite membrane has a very similar areal resistance to Nafion 212 even though the proton conductivity of the composite membrane is less than that of Nafion 212 due to thickness difference. It is presumed that they could lead to very similar fuel cell performance. Figure 8 shows I–V polarization curves of the MEAs using Nafion 212 and the composite membrane. First of all, it was found that the crossover of hydrogen in the MEA using the GA/#3_30-treated composite membrane exhibited approximately 0.5 mA cm^{-2} which confirms no defect composite membrane. As expected, the I–V polarization curves of two MEAs in activation and Ohmic region are very identical, but higher mass transport limitation is observed for Nafion 212. The main reason to show higher mass transport voltage loss at high current densities is due to water flooding at cathode since a water forming oxygen reduction reaction occurs as well as proton transport causes electro-osmotic drag from the anode to the cathode [40]. Under the same conditions except for the type of electrolyte membranes in the I–V polarization curves, electrolyte membranes could substantially affect the cathode flooding. The composite membrane facilitates back diffusion of water to prevail on electro-osmotic drag leading to the net water transport toward the anode [41]. Diffusion is driven by a gradient in concentration over a moving distance. Hence, the gradient becomes higher if the moving distance, i.e., the thickness of membranes, is shorter, leading to an increase in diffusion flux. In the same way, the composite membrane with less thickness than Nafion 212 would result in higher back diffusion of water and mitigate cathode flooding. Finally, it leads to lessening the mass transport voltage loss.

Table 7. Physical properties of Nafion 212 and the composite membrane using the GA/#3_30 and the summary of the I–V polarization curves of the membrane-electrode assemblies using Nafion 212 and the composite membrane.

Samples	Thickness (μm)	Proton Conductivity (S/cm)	Areal Resistance (Ω cm^2)	Ohmic Resistance from I–V (Ω cm^2)	High-Frequency Resistance (Ω cm^2)
Nafion 212	50	0.12	0.042	0.212	0.382
Composite membrane	27	0.069	0.039	0.233	0.409

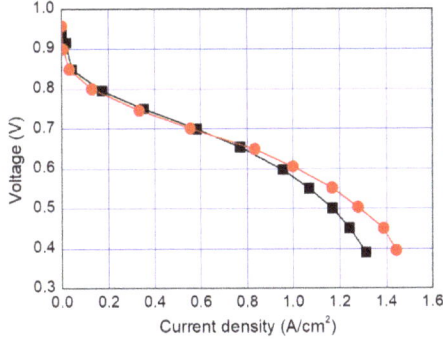

Figure 8. I–V polarization curves of the membrane-electrode assemblies using Nafion 212 (**black**) and the composite membrane (**red**) using the GA/#3_30-treated porous substrate.

4. Conclusions

In this study, we have developed a new hydrophilic treatment method for PTFE porous substrates to impregnate PFSA ionomers into strongly hydrophobic porous substrates. Two different PFTE substrates were used: the one has the thickness of ~5 μm and the porosity of ~80% and another has the thickness of ~25 μm and the porosity of ~60%. For hydrophilic treatment, we have used in-situ biomimetic silicification which the pyrogallol moiety in gallic acid (GA) with amino-terminated substances (ATS) such as siloxane generated a similar mussel-inspired adhesive coating via Michael addition/Schiff base reactions in alkaline conditions. We investigated three different ATS materials, i.e., 3-aminopropyltriethoxysilane (APTES), N-[3-(trimethoxysilyl)propyl]ethylenediamine (TMPEDA), and (3-trimethoxysilylpropyl)diethylenetriamine (TMPDETA) with GA. It was found that GA/APTES showed no hydrophilic treatment on both substrates with higher porosity than microfiltration or ultrafiltration membranes. On the other hand, GA/TMPEDA and GA/TMPDETA using ATS with the longer amino moieties showed effective hydrophilic treatment on both substrates. However, GA/TMPDETA has obtained the best contact angle result at less incubation time for the thinner substrates. The thinner substrate having ~5 μm treated by the GA/TMPDETA solution with the incubation time of 30 min shows the best hydrophilic treatment result. In addition, the composite membranes using the porous substrates show the highest proton conductivity and the lowest water uptake and swelling ratio. MEAs using the composite membranes (thinner and lower proton conductivity) and Nafion 212 (thicker and higher proton conductivity), which have similar areal resistance, are compared in I–V polarization curves. The I–V polarization curves of two MEAs in activation and Ohmic region are very identical. However, higher mass transport limitation is observed for Nafion 212 since the composite membrane with less thickness than Nafion 212 would result in higher back diffusion of water and mitigate cathode flooding. It could be concluded that the composite membrane would be advantageous in proton exchange membrane fuel cell application since it has similar areal resistance to Nafion 212 to obtain similar Ohmic loss and less thickness than Nafion 212 to allow higher water flux from cathode to anode to obtain lower mass transport loss in I–V polarization curves.

Author Contributions: Conceptualization, J.-S.P.; methodology, S.L. and J.-S.P.; experimentation, S.L. and J.-S.P.; validation, J.-S.P.; investigation, S.L. and J.-S.P.; resources, J.-S.P.; writing—original draft preparation, S.L. and J.-S.P.; writing—review and editing, J.-S.P.; supervision, J.-S.P.; project administration, J.-S.P.; funding acquisition, J.-S.P. All authors have read and agreed to the published version of the manuscript.

Funding: This research was funded by a 2018 Research Grant from Sangmyung University.

Conflicts of Interest: The authors declare no conflict of interest.

References

1. Peng, L.; Wei, Z. Catalyst engineering for electrochemical energy conversion from water to water: Water electrolysis and the hydrogen fuel cell. *Engineering* **2020**, *6*, 653–679. [CrossRef]
2. The Future of Hydrogen: Seizing Today's Opportunities. Available online: https://www.iea.org/reports/the-future-of-hydrogen (accessed on 15 October 2020).
3. Chaube, A.; Chapman, A.; Shigetomi, Y.; Huff, K.; Stubbins, J. The role of hydrogen in achieving long term Japanese energy system goals. *Energies* **2020**, *13*, 4539. [CrossRef]
4. Holladay, J.D.; Hu, J.; King, D.L.; Wang, Y. An overview of hydrogen production technologies. *Catal. Today* **2009**, *139*, 244–260. [CrossRef]
5. Zhang, W.Q.; Yu, B.; Chen, J.; Xu, J.M. Hydrogen production through solid oxide electrolysis at elevated temperatures. *Prog. Chem.* **2008**, *20*, 778–787.
6. Yu, H.; Yi, B. Hydrogen for energy storage and hydrogen production from electrolysis. *Strat. Stud. Chin. Acad. Eng.* **2018**, *20*, 58–65. [CrossRef]
7. Park, J.S.; Shin, M.S.; Kim, C.S. Proton exchange membranes for fuel cell operation at low relative humidity and intermediate temperature: An updated review. *Curr. Opin. Electrochem.* **2017**, *5*, 43–55. [CrossRef]
8. Uchida, M. PEFC catalyst layers: Effect of support microstructure on both distributions of Pt and ionomer and cell performance and durability. *Curr. Opin. Electrochem.* **2020**, *21*, 209–218. [CrossRef]
9. Pollet, B.G.; Kocha, S.S.; Staffell, I. Current status of automotive fuel cells for sustainable transport. *Curr. Opin. Electrochem.* **2019**, *16*, 90–95. [CrossRef]
10. Craig, S.G.; Kongkanand, A.; Masten, D.; Gu, W. Materials research and development focus areas for low cost automotive proton-exchange membrane fuel cells. *Curr. Opin. Electrochem.* **2019**, *18*, 81–89.
11. Song, C.H.; Park, J.S. Effect of Dispersion Solvents in Catalyst Inks on the Performance and Durability of Catalyst Layers in Proton Exchange Membrane Fuel Cells. *Energies* **2019**, *12*, 549. [CrossRef]
12. Liu, F.; Yi, B.; Xing, D.; Yu, J.; Zhang, H. Nafion/PTFE composite membranes for fuel cell applications. *J. Membr. Sci.* **2003**, *212*, 213–223. [CrossRef]
13. Tang, H.; Pan, H.; Wang, F.; Shen, P.K.; Jiang, S.P. Highly durable proton exchange membranes for low temperature fuel cells. *J. Phys. Chem.* **2007**, *111*, 8684–8690. [CrossRef] [PubMed]
14. Li, M.Q.; Scott, K. A polymer electrolyte membrane for high temperature fuel cells to fit vehicle applications. *Electrochim. Acta* **2010**, *55*, 2123–2128. [CrossRef]
15. Zhao, Y.; Yu, H.; Xing, D.; Lu, W.; Shao, Z.; Yi, B. Preparation and characterization of PTFE based composite anion exchange membranes for alkaline fuel cells. *J. Membr. Sci.* **2012**, *421*, 311–317. [CrossRef]

16. Wang, L.; Yi, B.L.; Zhang, H.M.; Liu, Y.H.; Xing, D.M.; Shao, Z.G.; Cai, Y.H. Sulfonated polyimide/PTFE reinforced membrane for PEMFCs. *J. Power Sources* **2007**, *167*, 47–52. [CrossRef]
17. Kim, K.H.; Ahn, S.Y.; Oh, I.H.; Ha, H.Y.; Hong, S.A.; Kim, M.S.; Lee, Y.K.; Lee, Y.C. Characteristics of the Nafion®-impregnated polycarbonate composite membranes for PEMFCs. *Electrochim. Acta* **2004**, *50*, 577–581. [CrossRef]
18. Rodgers, M.P.; Berring, J.; Holdcroft, S.; Shi, Z. The effect of spatial confinement of Nafion in porous membranes on macroscopic properties of the membrane. *J. Membr. Sci.* **2008**, *321*, 100–113. [CrossRef]
19. Tezuka, T.; Tadanaga, K.; Matsuda, A.; Hayashi, A.; Tatsumisago, M. Utilization of glass paper as a support of proton inorganicorganic hybrid 3-glycidoxypropyltrimethoxysilane. *Electrochem. Commun.* **2005**, *7*, 245–248. [CrossRef]
20. Lee, J.R.; Kim, N.Y.; Lee, M.S.; Lee, S.Y. SiO$_2$-coated polyimide nonwoven/Nafion composite membranes for proton exchange membrane fuel cells. *J. Membr. Sci.* **2011**, *367*, 265–272. [CrossRef]
21. Miyahara, T.; Hayano, T.; Matsuno, S.; Watanabe, M.; Miyatake, K. Sulfonated polybenzophenone/poly(arylene ether) block copolymer membranes for fuel cell applications. *ACS Appl. Mater. Interfaces* **2012**, *46*, 2881–2884. [CrossRef]
22. Goto, K.; Rozhanskii, I.; Yamakawa, Y.; Otsuki, T.; Naito, Y. Development of aromatic polymer electrolyte membrane with high conductivity and durability for fuel cell. *Polym. J.* **2009**, *41*, 95–104. [CrossRef]
23. Gubler, L.; Nauser, T.; Coms, F.D.; Lai, Y.H.; Gittleman, C.S. Prospects for durable hydrocarbon-based fuel cell membranes. *J. Electrochem. Soc.* **2018**, *165*, 3100–3103. [CrossRef]
24. Nouel, K.M.; Fedkiw, P.S. Nafion (R)-based composite polymer electrolyte membranes. *Electrochim. Acta* **1998**, *43*, 2381–2387. [CrossRef]
25. Yamaguchi, T.; Miyata, F.; Nakao, S. Polymer electrolyte membranes with a pore-filling structure for a direct methanol fuel cell. *Adv. Mater.* **2003**, *15*, 1198–1201. [CrossRef]
26. Penner, R.M.; Martin, C.R. Ion transporting composite membranes: I. Nafion-impregnated Gore-Tex. *J. Electrochem. Soc.* **1985**, *132*, 514–515. [CrossRef]
27. Satterfield, M.B.; Majsztrik, P.W.; Ota, H.; Benziger, J.B.; Bocarsly, A.B. In Mechanical properties of Nafion and Nafion/titania membranes for PEM fuel cells. In Proceedings of the 2006 AIChE Annual Meeting, San Francisco, CA, USA, 16 November 2006.
28. Shin, S.H.; Nur, P.J.; Kodir, A.; Kwak, D.H.; Lee, H.; Shin, D.; Bae, B. Improving the mechanical durability of short-side-chain perfluorinated polymer electrolyte membranes by annealing and physical reinforcement. *ACS Omega* **2019**, *4*, 19153–19163. [CrossRef]
29. Liu, Y.; Naguyen, T.; Kristian, N.; Yu, Y.; Wang, X. Reinforced and self-humidifying composite membrane for fuel cell applications. *J. Membr. Sci.* **2009**, *330*, 357–362. [CrossRef]
30. Cho, E.H.; Cheong, S.I.; Rhim, J.W. Study on the fouling reduction of the RO membrane by the coating with an anionic polymer. *Membranes* **2012**, *22*, 481–488.
31. Chaudhuri, R.G.; Sunayana, S.; Paria, S. Wettability of a PTFE surface by cationic–non-ionic surfactant mixtures in the presence of electrolytes. *Soft Matter* **2012**, *8*, 5429–5433. [CrossRef]
32. Eykens, L.; Sitter, K.D.; Dotremont, C.; Schepper, W.D.; Pinoy, L.; Bruggen, B.V.D. Wetting resistance of commercial membrane distillation membranes in waste streams containing surfactants and oil. *Appl. Sci.* **2017**, *7*, 118. [CrossRef]
33. Yang, X.; Sun, H.; Pal, A.; Bai, Y.; Shao, L. Biomimetic silicification on membrane surface for highly efficient treatments of both oil-in-water emulsion and protein wastewater. *ACS Appl. Mater. Interfaces* **2018**, *10*, 29982–29991. [CrossRef] [PubMed]
34. Zhan, K.; Kim, C.; Sung, K.; Ejima, H.; Yoshie, N. Tunicate-inspired gallol polymers for underwater adhesive: A comparative study of ctechol and gallol. *Biomacromolecules* **2017**, *18*, 2959–2966. [CrossRef] [PubMed]
35. Cheng, X.Q.; Wang, Z.X.; Guo, J.; Ma, J.; Shao, L. Designing multifunctional coatings for cost-effectively sustainable water remediation. *ACS Sustain. Chem. Eng.* **2018**, *6*, 1881–1890. [CrossRef]
36. Cheng, X.Q.; Wang, Z.X.; Zhang, Y.; Zhang, Y.; Ma, J.; Shao, L. Bio-inspired loose nanofiltration membranes with optimized separation performance for antibiotics removals. *J. Membr. Sci.* **2018**, *554*, 385–394. [CrossRef]
37. Shin, M.S.; Lim, S.; Park, J.H.; Kim, H.J.; Chae, S.; Park, J.S. Thermally crosslinked and quaternized polybenzimidazole ionomer binders for solid alkaline fuel cells. *Int. J. Hydrog. Energy* **2020**, *45*, 11773–11783. [CrossRef]

38. Kim, D.H.; Choi, Y.E.; Park, J.S.; Kang, M.S. Capacitive deionization employing pore-filled cation-exchange membranes for energy-efficient removal of multivalent cations. *Electrochim. Acta* **2019**, *295*, 164–172. [CrossRef]
39. Seo, D.; Lee, J.; Park, S.; Rhee, J.; Choi, S.W.; Shul, Y.G. Investigation of MEA degradation in PEM fuel cell by on/off cyclic operation under different humid conditions. *Int. J. Hydrog. Energy* **2011**, *36*, 1828–1836. [CrossRef]
40. Ji, M.; Wei, Z. A review of water management in polymer electrolyte membrane fuel cells. *Energies* **2009**, *2*, 1057–1106. [CrossRef]
41. Nguyen, T.V.; White, R.E. A water and heat management model for proton-exchange-membrane fuel cells. *J. Electrochem. Soc.* **1993**, *140*, 2178–2186. [CrossRef]

Publisher's Note: MDPI stays neutral with regard to jurisdictional claims in published maps and institutional affiliations.

 © 2020 by the authors. Licensee MDPI, Basel, Switzerland. This article is an open access article distributed under the terms and conditions of the Creative Commons Attribution (CC BY) license (http://creativecommons.org/licenses/by/4.0/).

Article

KOH-doped Porous Polybenzimidazole Membranes for Solid Alkaline Fuel Cells

Jong-Hyeok Park and Jin-Soo Park *

Department of Green Chemical Engineering, College of Engineering, Sangmyung University, Cheonan 31066, Korea; sbq6358@gmail.com
* Correspondence: energy@smu.ac.kr; Tel.: +82-41-550-5315

Received: 3 December 2019; Accepted: 18 January 2020; Published: 21 January 2020

Abstract: In this study the preparation and properties of potassium hydroxide-doped meta-polybenzimidazole membranes with 20–30 μm thickness are reported as anion conducting polymer electrolyte for application in fuel cells. Dibutyl phthalate as porogen forms an asymmetrically porous structure of membranes along thickness direction. One side of the membranes has a dense skin layer surface with 1.5–15 μm and the other side of the membranes has a porous one. It demonstrated that ion conductivity of the potassium hydroxide-doped porous membrane with the porogen content of 47 wt.% (0.090 S cm^{-1}), is 1.4 times higher than the potassium hydroxide-doped dense membrane (0.065 S cm^{-1}). This is because the porous membrane allows 1.4 times higher potassium hydroxide uptake than dense membranes. Tensile strength and elongation studies confirm that doping by simply immersing membranes in potassium hydroxide solutions was sufficient to fill in the inner pores. The membrane-electrode assembly using the asymmetrically porous membrane with 1.4 times higher ionic conductivity than the dense non-doped polybenzimidazole (*m*PBI) membrane showed 1.25 times higher peak power density.

Keywords: polybenzimidazole; solid alkaline fuel cell; asymmetrically porous film; porogen

---

1. Introduction

Fuel cells are considered to be low-pollution and high-efficiency energy conversion technology which directly convert fuel energy into electricity. Proton exchange membrane fuel cells (PEMFCs) are currently used in the niche energy market due to intensive research and development of proton exchange membranes, especially perfluorinated sulfonic acid ionomers. Solid alkaline fuel cells (SAFCs) are fuel cell technology using anion exchange membranes for the conduction of hydroxide ions from cathode to anode. Much attention has been recently received due to several advantages such as enhanced oxygen reduction in an alkaline environment, less expensive membrane materials, and easier water management. The recent review paper reported very good peak power density over 1 W/cm^2 at 0.5–0.7 V for pure hydrogen as fuel and a Pt-based catalyst [1]. Nevertheless, anion exchange membranes still suffer from lower ionic conductivity than proton exchange membranes, since the electrical mobility of hydroxide ions is only 56% of that of protons In addition, the review found that the durability of most of the cells reported was limited to less than 1000 h [1]. The main reason for the poor performance stability of SAFCs is due to the chemical degradation of cationic functional groups of anion exchange membranes. Anion conductive polymers must exhibit high chemical, mechanical and thermal stability, as well as good ion conduction ability. Most of the functional groups of anion conductive polymers used are quaternized ammonium groups, which show low chemical and thermal stability due to nucleophilic displacement or Hoffman elimination in alkaline environments above 80 °C [2,3]. Hence, anion-conducting ionomers with no positively charged functional groups weak to alkaline environments and with additional function to increase hydroxide ion conductivity are highly recommended for SAFCs [4–11].

Polybenzimidazole (PBI) shows unique amphoteric characteristic and is widely used in applications requiring excellent thermo-mechanical stability. Acid-doping (mostly phosphoric acid) PBI is commonly used in proton exchange membranes for fuel cells [12,13]. A heterocyclic benzimidazole ring with amphoteric characteristic could, however, protonate and deprotonate in acidic and basic environments, respectively. Hence, PBI can be equilibrated in aqueous solutions of alkali metal hydroxides to obtain anion conduction. In other words, (-N=) and (-NH-) of imidazole groups can be used as ion conductors by reacting with acids or bases. Potassium hydroxide (KOH)-doped PBI is used as anion conductive polymer where an alkaline solution in alkaline direct alcohol fuel cells or water electrolyzers is used as fuel. PBI satisfies the aforementioned recommendations for anion exchange membranes, such as no positively charged functional group and function of anion conduction [14–21]. However, during SAFC operations using pure hydrogen and air, the hydroxide ions doped in PBI might be reduced directly leading to a decrease in ionic conductivity. In addition, the leakage of doped hydroxide ions is inevitable due to the water generated at the anode and moved toward the cathode during the operation of SAFC [16].

In this study, asymmetrically porous PBI membranes (~30 μm) were prepared in order to obtain good hydroxide ion conductivity by containing hydroxide ions in pores along with hydroxide ions doped in PBI as well as to minimize leakage of hydroxide ions during SAFC operations by making a dense surface on one side of the asymmetrically porous membranes.

2. Materials

m-PBI (Dapazol®, MW 43 kDa) was purchased from Danish power system (DPS, Kvistgård, Denmark). N,N-dimethylacetamide (DMAc) as organic solvent and dibutyl phthalate as porogen were purchased from Sigma Aldrich Chemical Co. (St. Louis, MO, USA) and were used without further purification.

3. Experiment Methods

3.1. Preparation of Asymmetrically Porous PBI Membranes

Asymmetrically porous membranes using PBI were prepared using the following method: 2 g of PBI and 18 g of DMAc were added to a round three-necked flask and then mixed under nitrogen purging and a reflux at 120 °C for at least 12 h. Dibutyl phthalate as porogen was then added and mixed again for 3 h. The resulting mixed solution was cooled down to room temperature. The prepared mixed solution was stored for 12 h at ~3 °C and was then cast to be a layer with a thickness of 250 μm on a glass plate using a doctor blade. The coated glass plate was heated in a dry oven from 25 °C to 80 °C and then dried for 8 h at 120 °C. Finally, the solid film was completed in a vacuum for 12 h. A glass plate with a cast membrane was immersed in distilled water. The cast membrane was removed from the glass plate and immersed in methanol for 12 h and in a mixture of methanol/distilled water (1:1) for 12 h to completely remove porogen imbedded inside the membranes. The membrane was finally dried at 80 °C in a convective dry oven. Asymmetrically porous PBI membranes with various contents of porogen as summarized in Table 1 were prepared in the same manner.

3.2. Ion Conductivity and the Level of KOH Doping

The asymmetrically porous PBI membranes (P-PBI) prepared were doped by soaking in 6 M potassium hydroxide (KOH). The ionic conductivity of the KOH-doped membranes was measured at every 48-h interval in order to monitor the change in the level of doping of the membranes. The hydroxide ion conductivity of KOH-doped PBI membranes was measured in a clip cell system as illustrated in Figure 1 supplying an AC power source with a frequency range of 0.01 to 10^5 Hz and a voltage intensity of 10 mV using the potentiostat/galvanostat with a frequency response analyzer (Bio-Logics 150, Paris, France). The ion conductivity σ value of the membranes was obtained using the following formula with the impedance R value measured:

$$\sigma = \frac{t}{(R - R_{blank})A} \quad (1)$$

where σ is the hydroxide ion conductivity, t is the thickness of the membrane, R is the impedance of the polymer electrolyte membrane between two electrodes, R_{blank} is the impedance R of the KOH solution between two electrodes, and A is the surface area (cm^2) of the electrode in the clip cell [17].

Table 1. Composition for asymmetrically porous polybenzimidazole (PBI) membranes.

Sample Name	mPBI	Porogen (g)	Porogen Content (wt.%)
mPBI	2	0	0
P-PBI 41	2	1	41
P-PBI 44	2	1.6	44
P-PBI 47	2	1.8	47

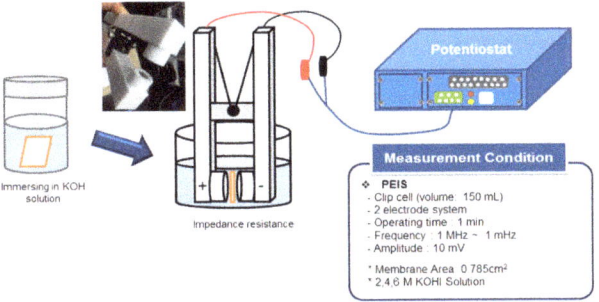

Figure 1. Clip cell for the measurement of impedance resistance.

The level of doping of KOH was measured to evaluate the effect on the actual conductivity. The KOH impregnation amount of the membrane was calculated using the following equation:

$$\text{KOH Uptake}(\%) = \frac{W_s - W_d}{W_d} \times 100 \quad (2)$$

where, W_s and W_d represent the weight of the wet and the dry membrane, respectively.

3.3. Characterization of KOH-doped Asymmetrically Porous PBI Membranes

The surface and cross-sectional images of the asymmetrically porous PBI membranes prepared were acquired by a field emission scanning electron microscope (SEM) (ULTRA PLUS, Carl Zeiss, Oberkochen, Germany) with an electron beam of 3 kV. The membranes were dried overnight in a vacuum oven at room temperature and then coated with gold (Au) using a sputter coater (Med 010, Oerlikon Balzers, Balzers, Liechtenstein) to achieve a specific coating thickness (~10 nm). SEM images were analyzed in terms of pore size and distribution using the image J program. The cast membranes were analyzed using an attenuated total reflectance (ATR) unit of Fourier transform infrared (FTIR) (/Raman spectrometer (IFS 66/S, Bruker Optik GmbH, Ettlingen, Germany).

The thermal stability of KOH-doped asymmetrically porous PBI membranes was evaluated using thermogravimetric analysis (TGA) (TGA 2050 CE, TA instruments, Inc., New Castle, DE, USA) at a rate of 10 °C min^{-1} from room temperature to 800 °C under a nitrogen atmosphere.

Tensile strength tests of un- and doped membranes using a micro material tester (Model 5848, Instron, Norwood, MA, USA) were measured to evaluate mechanical stability at 5.00 mm min^{-1} at ambient temperature and humidity. The sample shape was rectangular with a width of 10 mm and a length of 30 mm. All other procedures were based on the ASTM D882 standard test method.

3.4. Fabrication and Evaluation of Membrane-Electrode Assembly

For the electrochemical analyses of the asymmetrically porous PBI membranes prepared, a membrane-electrode assembly (MEA) was prepared by gas diffusion layer electrodes (GDE) using commercially available gas diffusion layers (23BC, SIGRACET®, SGL Carbon, Wiesbaden, Germany). The electrode has the effective area of 9 cm^2 with 0.4 mg of Pt/cm^2 for each electrode. A pair of GDEs and the KOH-doped membrane were assembled with the clamping pressure of the cell of 50 kgf. The dense surface of the asymmetrically porous membranes was faced to the anode for all MEAs, unless specifically stated. MEA performance tests were performed using KOH-doped GDE prepared by immersion in 6 M KOH for 24 h on both sides of alkali-doped asymmetrically porous PBI membrane. The performance and characterization of the MEAs were compared by measuring I-V polarization curves and impedance spectroscopy for high frequency resistance (HFR). The station conditions for the performance analysis of the MEAs were set at 60 °C with 0.3 L min^{-1} of hydrogen and oxygen to the anode and cathode, respectively.

4. Results and Discussion

PBI-based polymers have high thermal stability and mechanical strength, as well as excellent chemical resistance to acids and bases. In this study, thin asymmetrically porous PBI membranes with 20~30 μm thickness were prepared and used as an anion conducting membrane by doping KOH. Figure 2 shows photographs of non-porous and asymmetrically porous PBI membranes. The thin PBI membranes (on the far left hand side in Figure 2) are transparent since they have no pores. However, all of the P-PBI membranes are opaque due to light scattering by pores inside the membranes. The different colors of the upside and downside are associated with the asymmetric structure of the membranes. One side of the membranes (upside in Figure 2) is dense, while another side of the membranes (downside in Figure 2) has a porous surface. This structural difference can be confirmed in Figure 3. Figure 3 shows SEM images of both sides of the surface of the membranes of *m*PBI and P-PBI 41, 44, and 47. The upside and downside of the membranes show dense and porous surfaces, respectively. As the content of porogen increases, the number and size of pores in the downside surface of the membranes increase. As shown in the cross-sectional images of the membranes of Figure 3, all of the upside surfaces have dense skin layers, and pores are formed beneath the skin layers. The images of the downside surface show a few tiny pores in Figure 3a, but it seems that the formation of pores inside the membranes is not clear. Thus, the porogen content of 41 wt.% could be considered as the minimum amount to start forming pores inside the membranes. Table 2 summarizes the structural properties of the P-PBI membranes. It is inferred that the membrane drying results in an asymmetrically porous structure of the P-PBI membranes, since the thickness of the skin layers of the P-PBI membranes is inversely proportional to the content of porogen.

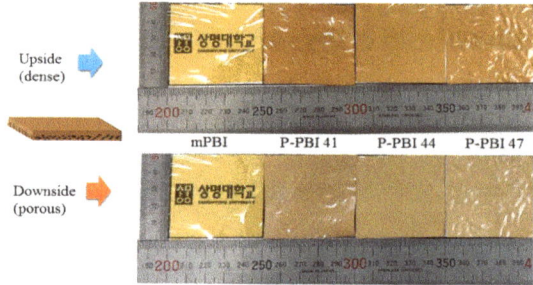

Figure 2. Photographs of non-porous and asymmetrically porous PBI membranes.

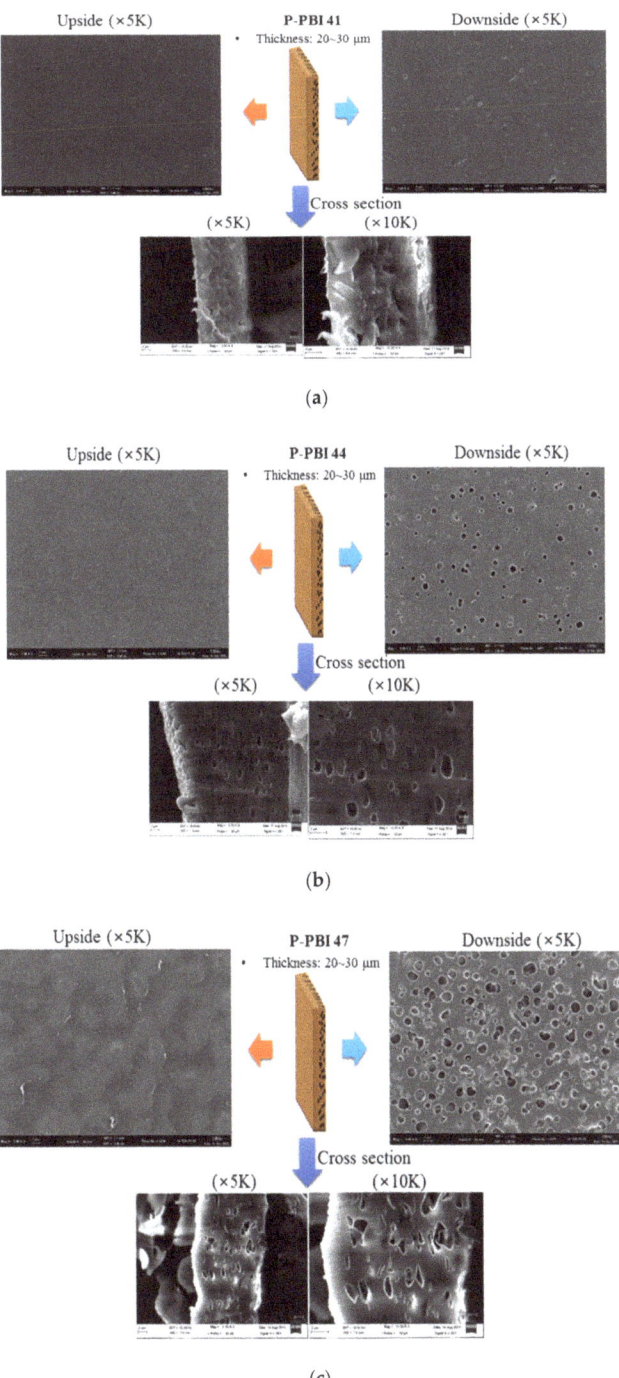

Figure 3. Photographs of non-porous and asymmetrically porous PBI membranes: (**a**) P-PBI 41, (**b**) P-PBI 44 and (**c**) P-PBI 47.

Table 2. Summary of the structural properties of the P-PBI membranes.

Sample Name	Average Pore Size (μm)	Range of Pore Size (μm)	Thickness of a Skin Layer
P-PBI 41	0.2	0.055–0.44	10–15
P-PBI 44	0.94	0.2–2	1.8–2
P-PBI 47	1.72	0.4–3.1	1.5

Chemical stability of the membranes in an alkaline environment during their KOH doping in a 6 M KOH solution was investigated. In Figure 4, the ATR-FTIR spectra of the KOH-doped membranes are shown, along with that of the non-doped PBI membrane (*m*PBI) as a reference. The *m*PBI membrane shows the peaks of C-N stretching at 1600 cm^{-1}, N-H at 1544 cm^{-1}, and imidazole group at 1284 cm^{-1}. After KOH doping, the adsorption peaks appear around 1625, 1510, and 1120 cm^{-1} due to the adsorption of water to O-H stretching, N-K deformation, and N-K plane bending, respectively. Compared with non-doped *m*PBI, the peak at 1284 cm^{-1} disappears and the peaks at 1510 cm^{-1} and 1120 cm^{-1} appear instead. This confirms that the group of imidazole reacts with the cation K$^+$ to contain the KOH in the membranes during KOH doping [14,17]. The P-PBI membranes show the same peaks to those of *m*PBI due to KOH doping inside polymer as well as pores.

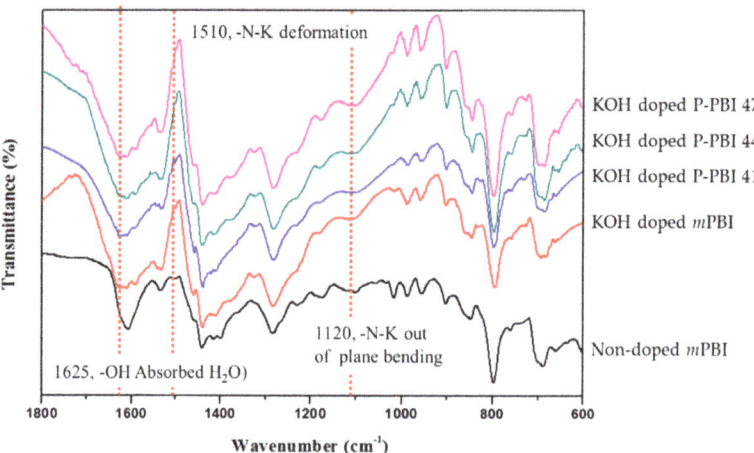

Figure 4. ATR-FTIR spectra of potassium hydroxide (KOH) doped P-PBI membranes.

The thermal and mechanical stability of the P-PBI membranes were investigated, and we tried to determine whether the porous structure weakens their mechanical properties. Figure 5 shows the TGA spectra of the non-doped and KOH-doped *m*PBI membranes and the KOH-doped P-PBI membranes. A decrease in weight up to 150 °C for all spectra is associated with dehydration of the membranes since the polybenzimidazole polymer is very hygroscopic and the KOH-doped membranes contain water absorbed during doping. No further decrease in weight up to 600 °C is shown and then the weight loss above 600 °C is associated with the decomposition of the polymer backbone [17]. The KOH-doped *m*PBI membrane shows less thermal stability than the non-doped mPBI membranes due to the plasticizing effect of KOH. Interestingly, the KOH-doped P-PBI membranes show very similar thermal stability to the KOH-doped *m*PBI membrane. This confirms that no KOH absorbed inside pores of the membranes is evaporated and no pores are collapsed up to 800 °C. Table 3 summarizes the results of the mechanical property tests on the non-doped and doped membranes to confirm the mechanical stability of the KOH-doped asymmetrically porous membranes. The non-doped membranes show much higher tensile strength than doped membranes, due to the plasticizing effect of KOH. As the content of porogen for both the non-doped and doped membranes increases, the tensile strength decreases due to porous structure [22]. For elongation, the doped membranes obtain higher values than the non-doped membranes due to an increase in softness. The elongation of the non-doped membranes increases as the content of porogen increases. However, the elongation of the doped membranes shows the opposite trend. Pores in the non-doped membranes play a role in increasing ductility of the polymer in good agreement with the previous study [23]. However, pores in the plasticized polymer for the doped membranes show a brittle response. It might be concluded that empty pores show a response reminiscent of an elastic–plastic material to result in KOH doping which fills up pores diminishes the effect of porosity. It infers that during KOH doping by simply immersing the P-PBI membranes a KOH solution is successfully filled in pores.

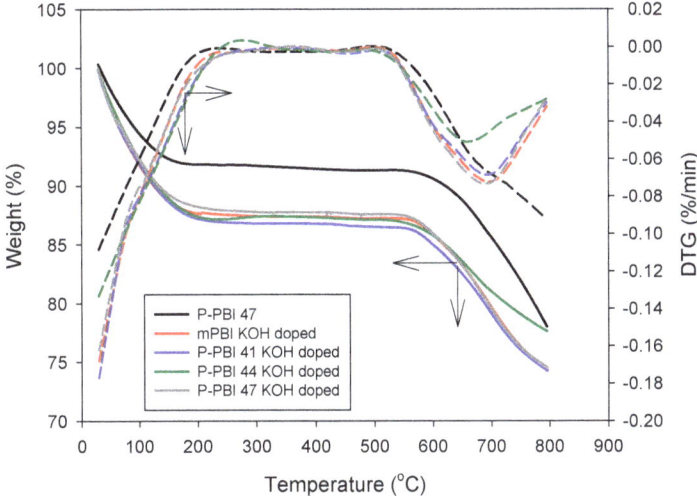

Figure 5. TGA spectra of KOH-doped P-PBI membranes.

Table 3. Properties of mechanical strength of KOH-doped *m*PBI and P-PBI membranes.

Sample Name	Non-Doped		6 M KOH-doped	
	Tensile Strength (MPa)	Elongation at Break (%)	Tensile Strength (MPa)	Elongation at Break (%)
*m*PBI	62.4 ± 1.2	6.62 ± 2.9	4.16 ± 0.65	80.6 ± 11.2
P-PBI 41	32.8 ± 2.7	8.10 ± 2.2	3.45 ± 0.46	52.9 ± 13.9
P-PBI 44	30.0 ± 0.8	10.6 ± 1.9	2.82 ± 0.34	54.3 ± 14.9
P-PBI 47	25.1 ± 2.6	11.8 ± 2.4	2.68 ± 0.54	25.8 ± 8.70

KOH doping of PBI polymer uses the reaction of (-N=) and (-NH-) with K^+ in PBI's Imidazole groups. The first mechanism is that *m*PBI, which is weakly basic due to the presence of -NH-, condenses with K^+ to generate water and hydrogen bonds. Thereafter, other excess KOH and water react with (-N=) to form an ionic bond with K^+. *m*PBI backbone is impregnated with a large amount of water, thereby forming ion clusters [24]. Figure 6 shows ion conductivity and KOH uptake of the membranes as a function of the content of porogen. Ion conductivity of the KOH-doped membranes shows a tendency to increase, even though ion conductivity of the P-PBI 44 (0.069 S cm^{-1}) slightly decreases, rather than the P-PBI 41 (0.075 S cm^{-1}), but higher than the *m*PBI (0.065 S cm^{-1}). Ion conductivity of the P-PBI 47 (0.090 S cm^{-1}) significantly increases 1.4 times higher than the *m*PBI. Nonetheless, the P-PBI 47 shows the highest ion conductivity due to surpassed KOH doping amount. This could be confirmed by KOH uptake as a function of the content of porogen. KOH uptake greatly increases as the content of porogen increases. Thus, ion conductivity should increase the similar tendency of the increase in ion conductivity of the membranes. However, ion conductivity of the P-PBI 44 behaves in an opposite way. It can be inferred that the formation position or distribution of pores formed inside the membranes is very significant. In other words, similarly to the general phenomenon in which the well-formation of an ion cluster network (e.g., hydrophilic/hydrophobic segregation) in swelled ion conducting membranes is significant to show high ion conductivity, the P-PBI membranes might fail to have well-connected pores doped by KOH which cause lower ionic conductivity.

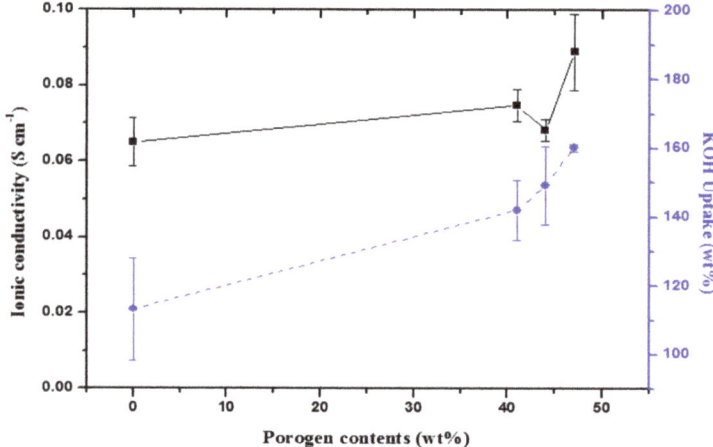

Figure 6. Ionic conductivity and KOH uptake of P-PBI membranes doped in 6 M KOH as a function of the content of porogen.

Table 4 summarizes the results of the current-voltage (I-V) polarization measurement of membrane-electrode assemblies (MEAs) using the KOH-doped membranes. Hydrogen is oxidized at the anode and oxygen is reduced at the cathode. Oxygen reduction generates hydroxide ions which migrate through the membranes towards the anode. The main two reactions are as follows:

At anode: $H_2 + 2OH^- \rightarrow 2H_2O + 2e^-$

At cathode: $O_2 + 2H_2O + 4e^- \rightarrow 4OH^-$

Open circuit voltage (OCV) of all the MEAs measured in this study exceeds 0.9 V, and thus the asymmetrically porous structure of the P-PBI membranes causes no gas crossover problem so as to prepare MEAs using the P-PBI membranes. There is no proportional relationship between the porogen content and OCV. Interestingly, the KOH-doped *m*PBI shows higher current density at high voltage (>0.55 V) than the P-PBI membranes, but the performance of the P-PBI membranes surpasses that of the *m*PBI membrane at low voltage (<0.55 V). It is associated with the characteristics of MEAs using the P-PBI membranes. The P-PBI membranes contains more KOH uptake, and KOH could be leaked out when the P-PBI membranes are compressed during unit cell assembly. Thus, KOH could fill in electrodes of MEAs to result in higher activation loss. Eventually, this affects the performance of the MEAs at voltages greater than 0.55 V. The performance of the MEAs below 0.55 V could be surpassed, since KOH filling electrodes could be ejected out of electrodes or absorbed towards the membranes, thereby resulting in decreased Ohmic loss in the MEA. In addition, the P-PBI membranes show higher ionic conductivity than the *m*PBI membranes. It is thought that the P-PBI membranes cause no serious problems in obtaining good MEA performance. Figure 7 exhibits the significantly linear relationship between the hydroxide ion conductivity of membranes and maximum power density of MEAs. It is believed that ion conductivity of membranes plays a crucial role in determining the maximum power density of MEAs. This could be a good way to increase ionic conductivity of membranes by means of both chemical doping inside the polymer and physically contained in pores to increase the KOH uptake of the membranes. Among the P-PBI membranes, the P-PBI 44 membrane shows the lowest performance. This is due to the abnormal behavior of ionic conductivity as shown in Figure 6. The MEA using P-PBI 47_R shows similar performance at low current density and less performance at high current density. This is associated with the fact that KOH leakage from the porous surface of P-PBI 47 pronounces liquid (water generated and KOH leaked) accumulation at the anode at high current densities. Thus, this would be an advantage of asymmetrically porous membranes.

Table 4. Summary of the performance results from I-V curves of membrane-electrode assemblies using mPBI and P-PBI membranes.

Sample Name	OCV (V)	Current Density at 0.8 V (mA cm^{-2})	Current Density at 0.6 V (mA cm^{-2})	Current Density at 0.3 V (mA cm^{-2})	Maximum Power Density (mW cm^{-2})
mPBI	0.95	38.1	89.0	151	56.8
P-PBI 41	0.93	25.9	75.8	196	66.2
P-PBI 44	0.97	25.6	72.4	183	62.1
P-PBI 47	0.93	26.1	80.3	213	71.0
P-PBI 47_R *	0.94	26.4	78.2	194	65.8

* P-PBI 47_R means MEA where the dense surface of P-PBI 47 is faced to cathode.

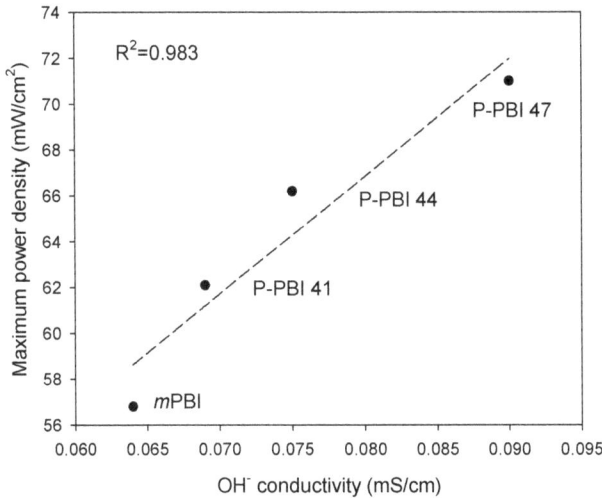

Figure 7. Correlation of maximum power density of MEAs with hydroxide ion conductivity of membranes.

5. Conclusions

Thin asymmetrically porous PBI membranes with 20–30 μm thickness were prepared using porogen and the membranes were doped by being immersed in a 6 M KOH solution to transform PBI membranes into hydroxide ion-conducting PBI membranes. The porous PBI membranes formed asymmetrical porous structures along the thickness direction during membrane drying. Eventually, one side of the membranes has a dense surface as skin layer, having 1.5–15 out of 20–30 μm total thickness, and the other side of the membranes has a porous one. This structure allows the membranes to achieve 1.4 times higher KOH uptake and ionic conductivity than dense mPBI membranes. One of the asymmetrically porous membranes behaved in an opposite way, since pore distribution formed inside the membrane was not well connected. Thus, it should be considered that the content of porogen is not only a parameter to prepare asymmetrically porous membranes with better ionic conductivity.

Mechanical property measurements confirmed that KOH doping by simply immersing membranes in KOH solutions was sufficient to fill in pores. If the asymmetrically porous membranes with higher KOH uptake by chemical doping inside polymer and physical containing in pores showed higher ionic conductivity, MEAs using the membranes showed higher MEA performance. The MEA performance at high voltage (>0.55 V in this study) was worse than that of MEAs using non-porous mPBI membranes, due to the higher activation loss, possibly cause by leaking of KOH out of the asymmetrically porous membranes. In addition, it was found that the performance is influenced by the position of the porous surface. It is a disadvantage to apply the MEAs for residential fuel cell systems normally operated at a fixed voltage in the range of 0.7–0.75 V. To overcome this problem, in a further study, researchers could develop membranes with skin layers on both sides of the membrane and a porous structure in between skin layers to prevent KOH leaking-out and to contain higher KOH chemically and physically.

Author Contributions: Conceptualization, J.-S.P.; methodology, J.-S.P.; experimentation, J.-H.P. and J.-S.P.; validation, J.-S.P.; investigation, J.-H.P. and J.-S.P.; resources, J.-S.P.; writing—original draft preparation, J.-H.P. and J.-S.P.; writing—review and editing, J.-S.P.; supervision, J.-S.P.; project administration, J.-S.P.; funding acquisition, J.-S.P. All authors have read and agreed to the published version of the manuscript.

Funding: This research was funded by a 2017 Research Grant from Sangmyung University.

Acknowledgments: This research was funded by a 2017 Research Grant from Sangmyung University.

Conflicts of Interest: The authors declare no conflict of interest.

References

1. Dekel, D.R. Review of cell performance in anion exchange membrane fuel cells. *J. Power Sources* **2018**, *375*, 158–169. [CrossRef]
2. Tomoi, M.; Yamaguchi, K.; Ando, R.; Kantake, Y.; Aosaki, Y.; Kubota, H. Synthesis and thermal stability of novel anion exchange resins with spacer chains. *J. Appl. Polym. Sci.* **1997**, *64*, 1161–1167. [CrossRef]
3. Henkensmeier, D.; Cho, H.-R.; Kim, H.-J.; Carolina, N.-K.; Leppin, J.; Dyck, A.; Jang, J.-H.; Cho, E.; Nam, S.-W.; Lim, T.-H. Polybenzimidazolium hydroxides—Structure, stability and degradation. *Polym. Degrad. Stab.* **2012**, *97*, 264–272. [CrossRef]
4. Wang, Y.-J.; Qiac, J.; Baker, R.; Zhang, J. Alkaline polymer electrolyte membranes for fuel cell applications. *Chem. Soc. Rev.* **2013**, *42*, 5768–5787. [CrossRef]
5. Couture, G.; Alaaeddine, A.; Boschet, F.; Ameduri, B. Polymeric materials as anion-exchange membranes for alkaline fuel cells. *Prog. Polym. Sci.* **2011**, *36*, 1521–1557. [CrossRef]
6. Hickner, M.-A.; Herring, A.-M.; Coughlin, E.-B. Anion exchange membranes: Current status and moving forward. *J. Polym. Sci. Polym. Phys.* **2013**, *51*, 1727–1735. [CrossRef]
7. Maurya, S.; Shin, S.-H.; Kim, Y.; Moon, S.-H. A review on recent developments of anion exchange membranes for fuel cells and redox flow batteries. *RSC Adv.* **2015**, *5*, 37206–37230. [CrossRef]
8. Park, S.-H.; Yim, S.-D.; Yoon, Y.-G.; Lee, W.-Y.; Kim, C.-S. Performance of solid alkaline fuel cells employing anion-exchange membranes. *J. Power Sources* **2008**, *178*, 620–626. [CrossRef]
9. Park, J.-S.; Park, G.-G.; Park, S.-H.; Yoon, Y.-G.; Kim, C.-S.; Lee, W.-Y. Development of solid-state alkaline electrolytes for solid alkaline fuel cells. *Macromol. Symp.* **2007**, *249–250*, 174–182. [CrossRef]
10. Choi, J.; Byun, Y.-J.; Lee, S.Y.; Jang, J.H.; Henkensmeier, D.; Yoo, S.J.; Hong, S.-A.; Kim, H.-J.; Sung, Y.-E.; Park, J.-S. Poly(arylene ether sulfone) with tetra(quaternary ammonium) moiety in the polymer repeating unit for application in solid alkaline exchange membrane fuel cells. *Int. J. Hydrog. Energy* **2014**, *39*, 21223–21230. [CrossRef]
11. Shin, M.-S.; Byun, Y.-J.; Choi, Y.-W.; Kang, M.-S.; Park, J.-S. On-site crosslinked quaternized poly(vinyl alcohol) as ionomer binder for solid alkaline fuel cells. *Int. J. Hydrog. Energy* **2014**, *39*, 16556–16561. [CrossRef]
12. Aili, D.; Jankova, K.; Han, J.; Bjerrum, N.J.; Jensen, J.O.; Li, Q. Understanding ternary poly(potassium benzimidazolide)-based polymer electrolytes. *Polymer* **2016**, *84*, 304–310. [CrossRef]
13. Hwang, K.; Kim, J.-H.; Kim, S.-Y.; Byun, H. Preparation of polybenzimidazole-based membranes and their potential applications in the fuel cell system. *Energies* **2014**, *7*, 1721–1732. [CrossRef]
14. Couto, R.-N.; Linares, J.-J. KOH-doped polybenzimidazole for alkaline direct glycerol fuel cells. *J. Membr. Sci.* **2015**, *486*, 239–247. [CrossRef]

15. Zeng, L.; Zhao, T.-S.; An, L.; Zhao, G.; Yan, X.-H. A high-performance sandwiched-porous polybenzimidazole membrane with enhanced alkaline retention for anion exchange membrane fuel cells. *Energy Environ. Sci.* **2015**, *8*, 2768–2774. [CrossRef]
16. Hou, H.; Sun, G.; He, R.; Sun, B.; Jin, W.; Liu, H.; Xin, Q. Alkali doped polybenzimidazole membrane for alkaline direct methanol fuel cell. *Int. J. Hydrog. Energy* **2008**, *33*, 7172–7176. [CrossRef]
17. Aili, D.; Jankova, K.; Li, Q.; Bjerrum, N.-J.; Jensen, J.-O. The stability of poly (2,2′-(m-phenylene)-5,5′-bibenzimidazole) membranes in aqueous potassium hydroxide. *J. Membr. Sci.* **2015**, *492*, 422–429. [CrossRef]
18. Wang, Y.; Wang, Q.; Wan, L.; Han, Y.; Hong, Y.; Huang, L.; Yang, X.; Wang, Y.; Zaghib, K.; Zhou, Z. KOH-doped polybenzimidazole membrane for direct hydrazine fuel cell. *J. Colloid Interface Sci.* **2020**, *563*, 27–32. [CrossRef]
19. Penchev, H.; Borisov, G.; Petkucheva, E.; Ublekov, F.; Sinigersky, V.; Radev, I.; Slavcheva, E. Highly KOH doped para-polybenzimidazole anion exchange membrane and its performance in Pt/Ti$_n$O$_{2n-1}$ catalyzed water electrolysis cell. *Mater. Lett.* **2018**, *221*, 128–130. [CrossRef]
20. Wu, Q.X.; Pan, Z.F.; An, L. Recent advances in alkali-doped polybenzimidazole membranes for fuel cell applications. *Renew. Sust. Energy Rev.* **2018**, *89*, 168–183. [CrossRef]
21. Konovalova, A.; Kim, H.; Kim, S.; Lim, A.; Park, H.S.; Kraglund, R.M.; Aili, D.; Jang, J.H.; Kim, H.-J.; Henkensmeier, D. Blend membranes of polybenizimidazole and an anion exchange ionomer (FAA3) for alkaline water electrolysis: Improved alkaline stability and conductivity. *J. Membr. Sci.* **2018**, *564*, 653–662. [CrossRef]
22. Zarrin, H.; Jiang, G.; Lam, G.-Y.; Fowler, M.; Chen, Z. High performance porous polybenzimidazole membrane for alkaline fuel cells. *Int. J. Hydrog. Energy* **2014**, *39*, 18405–18415. [CrossRef]
23. Yadav, S.; Saldana, C.; Murthy, T.G. Porosity and geometry control ductile to brittle deformation in indentation of porous solids. *Int. J. Solids Struct.* **2016**, *88–89*, 11–16. [CrossRef]
24. Zeng, L.; Zhao, T.-S.; An, L.; Zhao, G.; Yan, X.-H. Physicochemical properties of alkaline doped polybenzimidazole membranes for anion exchange membrane fuel cells. *J. Membr. Sci.* **2015**, *493*, 340–348. [CrossRef]

© 2020 by the authors. Licensee MDPI, Basel, Switzerland. This article is an open access article distributed under the terms and conditions of the Creative Commons Attribution (CC BY) license (http://creativecommons.org/licenses/by/4.0/).

Article

Pore-Filled Anion-Exchange Membranes with Double Cross-Linking Structure for Fuel Cells and Redox Flow Batteries

Do-Hyeong Kim and Moon-Sung Kang *

Department of Green Chemical Engineering, College of Engineering, Sangmyung University, Cheonan 31066, Korea; dohyeongkim665@gmail.com
* Correspondence: solar@smu.ac.kr; Tel.: +82-41-550-5383

Received: 19 August 2020; Accepted: 10 September 2020; Published: 11 September 2020

Abstract: In this work, high-performance pore-filled anion-exchange membranes (PFAEMs) with double cross-linking structures have been successfully developed for application to promising electrochemical energy conversion systems, such as alkaline direct liquid fuel cells (ADLFCs) and vanadium redox flow batteries (VRFBs). Specifically, two kinds of porous polytetrafluoroethylene (PTFE) substrates, with different hydrophilicities, were utilized for the membrane fabrication. The PTFE-based PFAEMs revealed, both excellent electrochemical characteristics, and chemical stability in harsh environments. It was proven that the use of a hydrophilic porous substrate is more desirable for the efficient power generation of ADLFCs, mainly owing to the facilitated transport of hydroxyl ions through the membrane, showing an excellent maximum power density of around 400 mW cm^{-2} at 60 °C. In the case of VRFB, however, the battery cell employing the hydrophobic PTFE-based PFAEM exhibited the highest energy efficiency (87%, cf. AMX = 82%) among the tested membranes, because the crossover rate of vanadium redox species through the membrane most significantly affects the VRFB efficiency. The results imply that the properties of a porous substrate for preparing the membranes should match the operating environment, for successful applications to electrochemical energy conversion processes.

Keywords: pore-filled anion-exchange membranes; double cross-linking structures; alkaline direct liquid fuel cells; vanadium redox flow batteries; porous PTFE substrates

1. Introduction

Ion-exchange membranes (IEMs), which can selectively transport counter ions having the opposite charge to the fixed charge groups, and connected to the membrane matrix by means of the Donnan exclusion, have been widely utilized in many desalination processes such as electrodialysis (ED) [1–3], diffusion dialysis (DD) [4,5], membrane capacitive deionization (MCDI) [6–8], etc. Recently, IEMs have also been successfully applied in several electrochemical energy production and storage systems, including fuel cells (FCs) [9–13], reverse electrodialysis (RED) [14,15], redox flow batteries (RFBs) [16–19] and so on. Particularly, proton-exchange membranes (PEMs), such as Nafion, have been widely utilized in energy conversion processes, owing to their excellent proton conductivity and chemical stability [20].

Among various energy conversion systems, the application to fuel cells has been the most actively researched. As is well known, there are several types of fuel cells, depending on the kinds of electrolytes used and operation conditions, including the proton-exchange membrane fuel cell (PEMFC), alkaline fuel cell (AFC), phosphoric acid fuel cell (PAFC), molten carbonate fuel cell (MCFC), solid oxide fuel cell (SOFC), and direct methanol fuel cell (DMFC) [21]. Recently, alkaline direct liquid fuel cells (ADLFCs), utilizing liquid fuels and anion-exchange membranes (AEMs), have also attracted a lot of interest as one of promising energy production systems [22]. Like traditional alkaline fuel cells, ADLFCs have several attractive features, such as operation at relatively low temperature, fast electrode reaction (oxygen reduction reaction (ORR) at cathode), use of nonprecious metal catalysts, low fuel crossover, easy water management, and so on [22–24]. In addition, as liquid fuels, alcohols such as methanol, ethanol, and glycerol have been the most widely studied [24]. The use of formate alkaline solutions as liquid fuels for ADLFCs has recently attracted much attention because it can realize efficient power sources for portable electronic devices [11,25]. They can provide several benefits over alcohols, e.g., fast oxidation kinetics, theoretically high cell potential and power densities, and low fuel crossover [25,26]. AEMs are one of the key components determining the energy conversion performances of ADLFCs. Therefore, various studies have been conducted to develop high-performance AEMs for successful applications to alkaline membrane fuel cells. In particular, structural studies on the polymer backbone and anion-exchange groups are being carried out [27,28]. In addition, the state-of-the-art commercial AEMs such as Tokuyama A201 and A901 exhibited excellent performance in alkaline energy conversion processes, including AFCs [28], alkaline water electrolysis [29], and alkali metal-air batteries [30]. For example, the AFCs employing Tokuyama A901 achieved high power density of around 450 mW cm^{-2} at specific conditions [28]. Unfortunately, however, the ion conductivity of AEMs is significantly lower compared to that of PEMs (e.g., Nafion), and the chemical stability should also be further improved for practical applications.

Meanwhile, RFBs are one of the prospective large-scale electricity storage technologies and require efficient membranes for separating different redox species [31]. Among many kinds of RFBs, all-vanadium redox flow batteries (VRFB), in which vanadium redox species are used as the active electrode materials, have been the most actively researched and utilized due to advantages, such as long battery cycle life, reduced cross-contamination, high electrochemical activity, and so on [32,33]. The ideal membranes for VRFB should possess low vanadium ion permeability to reduce self-discharge and achieve high coulomb efficiency (CE), low area resistance to decrease losses in voltage efficiency (VE), and excellent chemical stability [34]. However, even though perfluorinated PEMs such as Nafion have been widely used as a separator in VRFBs, owing to their high proton permeability and chemical stability, they suffer from the significant crossover of cationic redox species during operation, which can result in a decrease in the battery efficiency [32]. In addition, the employment of expensive perfluorinated PEMs such as Nafion could significantly elevate the system cost [33]. Among various approaches to solving this problem, the use of AEMs has received much attention due to their low permeability of cationic species by the Donnan exclusion effect, and potentially low membrane cost [33,34]. To be successfully used in VRFB, however, their poor chemical stability in harsh acidic environments, and relatively low ion conductivity should be further enhanced [33]. For example, the ion conductivity of Tokuyama A201 is shown to be about 42 mS cm^{-1} [29] which is much smaller than that of Nafion (86 mS cm^{-1} for Nafion 1035 [35]).

Among the various kinds of IEMs, pore-filled IEMs (PFIEMs), which are composed of a thin and mechanically strong porous substrate film, and polyelectrolyte that fills the pores, are known to provide both the high ion conductivity and mechanical strength [36–39]. They could also be fabricated by a simple and cheap manufacturing process [39]. Recently, we have developed novel pore-filled AEMs (PFAEMs) for application to electrochemical energy conversion systems [11]. The results demonstrated that the PFAEMs, consisting of a thin porous polytetrafluoroethylene (PTFE) substrate and highly cross-linked ionomer, could provide both excellent electrochemical properties and alkaline stability [11]. Unfortunately, however, the power generation was much poorer than those of the state-of-the-art such as the Tokuyama A901 [28].

In this work, we have dramatically improved the performances of the PFAEMs by choosing a proper porous substrate and adjusting the cross-linking degree for successful applications to ADLFC and VRFB. Namely, two different kinds of porous PTFE substrate, i.e., hydrophobic and hydrophilic grades, were chosen and utilized for the comparative study, and membrane cross-linking was also finely controlled. Moreover, we have systematically characterized the prepared membranes via various experimental analyses, including membrane-electrode assembly (MEA) tests, VRFB charge-discharge tests, Fenton oxidation, vanadium oxidative stability, and crossover rate evaluations, etc.

2. Materials and Methods

2.1. Materials and Membrane Preparation

Two different grades of highly porous PTFE film (i.e., hydrophobic and hydrophilic grades, Advantec MFS, Inc., Tokyo, Japan) were employed as the substrates for preparing PFAEMs. The specifications of the porous PTFE substrates are listed in Table 1, showing the almost identical pore dimensions. N,N'-Dimethylaminoethyl methacrylate (DMAEMA) and divinylbenzene (DVB) were chosen as the main monomer and cross-linker, respectively. p-Xylylene dichloride (XDC) was used for both the quaternization and cross-linking of the polymer, and benzophenone (BP) was selected as the initiator for the photo-polymerization. The reaction scheme of DMAEMA-DVB copolymer, which is cross-linked and quaternized by XDC, is suggested in Figure 1. As shown in the scheme, the synthesized ionomer has a double cross-linking structure by DVB and XDC, realizing a high cross-linking degree and ion-exchange capacity, at the same time. For the membrane preparation, first, monomer mixtures consisting of 79–98 wt% DMAEMA, 1–20 wt% DVB, and 1 wt% BP were prepared. Porous PTFE substrate was then dipped in the monomer solution for several hours, followed by a photo-induced polymerization in a lab-made ultraviolet (UV) chamber (high pressure, lamp power = 1 kW) for 10 min. For the successive quaternization and cross-linking, the PTFE substrate filled with poly(DMAEMA-DVB) was immersed in 0.05 M XDC-EtOH solution at 50 °C for 12 h and then sequentially treated with 0.5 M NH_4Cl, 0.5 M HCl, and distilled water. The fabricated PFAEMs were then immersed in 0.5 M NaCl or 1.0 M KOH before evaluation. All the reagents were supplied by Sigma-Aldrich (St. Louis, MO, USA) and utilized as received. We also chose Neosepta® AMX (Astom Corp., Tokyo, Japan) as the reference membrane to compare with the PFAEMs fabricated in this work.

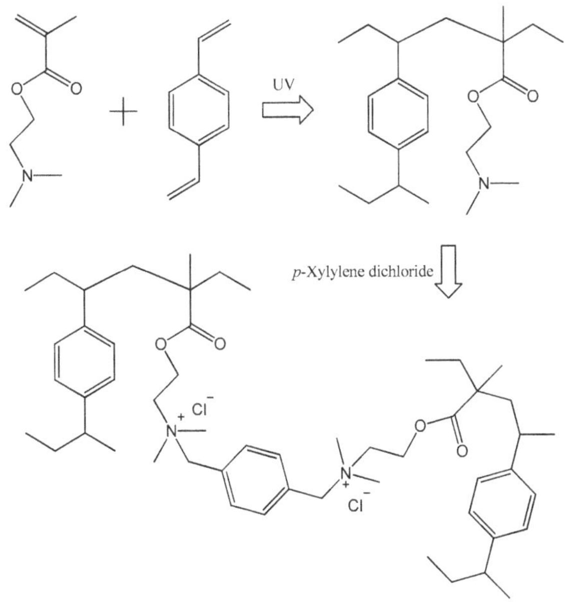

Figure 1. Reaction scheme of the anion-exchangeable polymer, with quaternary ammonium groups and two different kinds of cross-linking sites.

Table 1. Specifications of two porous polytetrafluoroethylene (PTFE) substrates utilized for preparing pore-filled anion-exchange membranes (PFAEMs) in this work.

Substrate	Thickness (µm)	Pore Size (µm)	Porosity (%)
Hydrophobic PTFE (T020A142C)	80	0.2	74
Hydrophilic PTFE (H020A142C)	35	0.2	71

2.2. Membrane Characterizations

The synthesis of the anion-exchangeable polymer was confirmed by FT-IR (FT/IR-4700, Jasco, Tokyo, Japan) analysis. The morphological characteristics of the porous substrates and prepared PFAEMs were investigated by employing field emission scanning electron microscopy (FE-SEM, TESCAN, Czech). The surface hydrophilicity of the membranes was also evaluated using a contact angle analyzer (Phoenix 150, SEO Co., Suwon-si, Korea). The water uptake (WU) of the membranes was determined using the following equation:

$$WU\ (\%) = \left(\frac{W_{wet} - W_{dry}}{W_{dry}}\right) \times 100 \tag{1}$$

where W_{dry} and W_{wet} are the dry and wet membrane weights, respectively. A traditional titration method was employed to determine the ion-exchange capacity (IEC) of the membranes. After the pre-equilibrium in 0.5 M NaCl, chloride ions in the membrane were fully exchanged with sulfate ions in 0.25 M Na$_2$SO$_4$. The amount of Cl$^-$ was then quantitatively analyzed by titration with an AgNO$_3$ standard solution, and the IEC values were calculated using the following equation:

$$IEC\ (meq./g_{dry\ memb}) = \frac{N_{Cl^-} \cdot V_s}{W_{dry}} \tag{2}$$

where N_{Cl^-} is the normal concentration of Cl$^-$ (meq./L), V_s is the solution volume (L), and W_{dry} is the dry membrane weights (g). Both the ion conductivity (σ) and electrical area resistance (EAR) of the membranes were evaluated in a 1.0 M KOH solution at room temperature using a lab-made clip cell and an LCZ meter. The σ values were obtained from the following equation:

$$\sigma\ (S/cm) = \frac{l}{R_{memb} \cdot A} \tag{3}$$

where R_{memb} is the resistance (Ω), l is the thickness (cm), and A is the effective area (cm^2) of the tested membrane. The EAR was estimated using the following equation:

$$EAR\ (\Omega\ cm^2) = \left(|Z|_{sample} \cdot \cos\theta_{sample} - |Z|_{blank} \cdot \cos\theta_{blank}\right) \cdot A \tag{4}$$

where $|Z|$ is the magnitude of impedance (Ω), θ is the phase angle, and A is the effective area (cm^2) of the tested membrane. The transport number (t_-) for counter ions (Cl$^-$) was obtained by measuring the cell potential (=electromotive force, emf) using a two-compartment cell (membrane area = 0.785 cm^2; each volume = 0.23 dm^3) equipped with a pair of Ag/AgCl reference electrodes. As a result, the t_- values were determined by the following equation:

$$E_m = \frac{RT}{F}(1 - 2t_-)\ln\frac{a_L}{a_H} \tag{5}$$

where E_m is the cell potential, F the Faraday constant, T the absolute temperature, R the molar gas constant, and a_H and a_L the activity in high and low concentration compartments, respectively. The alkaline stability of the membranes was evaluated by conventional soaking tests under a harsh alkaline environment (1 M KOH; 60 °C; 500 h). The oxidative stability of the membranes was also checked by soaking experiments using Fenton's reagent (3% H_2O_2 containing 2 ppm $FeSO_4$) at 80 °C for 6 h. The time-course changes in the transport number and membrane weight were recorded to evaluate the alkaline and oxidative stabilities, respectively. The fuel (hydrazine, N_2H_4) crossover rates through the membranes were estimated via conventional 2-compartment diffusion cell tests. The time-course change in N_2H_4 content in the low concentration compartment was quantitatively analyzed using UV/Vis spectroscopy (UV-2600i, Shimadzu Co., Kyoto, Japan). The diffusion coefficient (D) of N_2H_4 through a membrane was calculated by the following equation [16]:

$$D\,(\text{cm}^2/\text{s}) = \frac{dC_L}{dt}\frac{\delta V_L}{A(C_H - C_L)} \quad (6)$$

where t is the time, δ is the membrane thickness, V_L is the low concentration compartment volume, A is the membrane area, and C_H and C_L are the molar concentrations, in high and low concentration compartments, respectively. In addition, the overall dialysis coefficient (K_A) of vanadium cations through a membrane was determined using a two-compartment cell (effective area = 4 × 4 cm^2), filled with 1 M $VOSO_4$/2.0 M H_2SO_4 (feed) and 1 M $MgSO_4$/2.0 M H_2SO_4 (permeate). During the tests, the time-course change in vanadyl (VO^{2+}) ion concentration in the permeate compartment was recorded by measuring the solution absorbance using UV/Vis spectroscopy. The K_A values were determined from the dependence of the component concentration and volume changes upon time, using the following equation [40]:

$$\ln \frac{c_{A0}^I}{c_{A0}^I - \frac{1+k_V}{k_V}c_A^{II}} = \frac{1+k_V}{k_V}\frac{A}{V^{II}}K_A\tau \quad (7)$$

where c_{A0}^I is the initial molar concentration of component A in the feed compartment. c_A^I and c_A^{II} are the molar concentrations of component A in the feed (I) and permeate (II) compartments, respectively. A is the membrane area, τ is time, V^I and V^{II} are the solution volume in the feed (I) and permeate (II) compartments, respectively, and k_v is the solution volume ratio of both compartments (=V^I/V^{II}). In this work, the vanadium oxidative stability of the membranes was also confirmed. The membrane specimens (2 × 2 cm^2) were immersed in 0.1 M V_2O_5 solution (in 5 M H_2SO_4) and stored at 40 °C for about 200 h to evaluate the oxidative stability of membranes in a vanadium electrolyte solution. The time-course change in the oxidation state of vanadium ions was monitored by measuring the solution absorbance using UV/Vis spectroscopy [41].

2.3. MEA Performance Tests (ADLFC)

A lab-made Ni/C [2] and Pt/C (46.7%, Tanaka Co., Tokyo, Japan) were chosen as anode and cathode electrocatalysts, respectively. The electrocatalyst solutions were directly sprayed on the surface of the membranes (3 × 3 cm^2; OH-form), and the total loading amount of Ni and Pt was revealed to be about 2 mg cm^{-2} and 1 mg cm^{-2}, respectively. As a binder in the electrocatalyst inks, commercially available anion-exchange ionomer (AS-4, Tokuyama Co., Tokyo, Japan) was utilized. The catalyst-coated membrane (CCM) was inserted between two sheets of carbon fiber composite paper (TGP-H-060, Toray Inc., Tokyo, Japan) used as a gas diffusion layer (GDL). The CCM and GDLs were then assembled with a clamping pressure of about 5.5 MPa. The current–voltage (I–V) polarization characteristics of the prepared MEAs were investigated using a single cell, having an effective area of 9 cm^2 at 60 °C. Liquid fuel (4 M N_2H_4/4 M KOH) and humidified O_2 gas (of 100% relative humidity (RH)) were fed to the anode and cathode at the flow rate of 5 mL min^{-1} and 500 sccm, respectively. In addition, for the I–V polarization curve measurement, a current was stepped up by 0.01 A and then maintained for 30 s at each step to gain a stable response.

2.4. Charge-Discharge Tests (VRFB)

2 M $V_2(OSO_4)_3$ in 3 M H_2SO_4 (as anolyte) and 2 M $VOSO_4$ in 3 M H_2SO_4 (as catholyte) were employed as the electrolyte solutions to evaluate the charge–discharge performance in the VRFB

systems utilizing various membranes. The galvanostatic charge–discharge experiments were performed utilizing a lab-made RFB cell (membrane area = 12.5 cm^2) containing a pair of carbon felt electrodes (GF20-3, Nippon Graphite, Otsu-shi, Japan) with a battery cycler (WBCS3000S, Wonatech, Seoul, Korea) in the potential range of 0.9–1.9 V at 0.25 A. All the tests were carried out at room temperature.

3. Results and Discussion

Figure 2 shows the FT-IR spectra of the porous PTFE substrate and prepared membranes. In the spectra of both the base membrane (PS + Poly(DMAEMA-DVB)) and PFAEM (PS + Poly (DMAEMA-DVB-XDC)), the absorption bands at 1726 cm^{-1} and 1456 cm^{-1}, which are assigned to the stretch vibration of carbonyl group (C=O), and the C=C in plane stretch vibration of the benzene ring, respectively, indicating the successful in situ synthesis of poly(DMAMEA-DVB) inside the pores of the PTFE substrate [42]. In addition, the absorption band at around 3400 cm^{-1}, which is assigned to the stretching vibration of the N-H$^+$ group clearly elucidates the introduction of quaternary ammonium sites into the membrane [42].

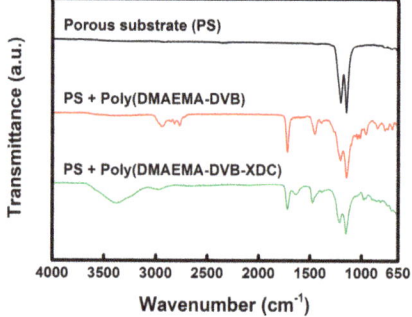

Figure 2. FT-IR spectra of porous PTFE substrate and pore-filled membranes.

The FE-SEM analyses were carried out to investigate the morphology and pore-filling of the membranes and the images are shown in Figure 3. The cross-sectional FE-SEM images of the PTFE substrate films show a highly porous structure, and the pores in the substrates were revealed to be completely filled with ionomer after the membrane fabrication. In addition, nano-sized metal oxide particles (e.g., Al$_2$O$_3$) were observed in the images of hydrophilic PTFE-based samples (Figure 3c,d), which might enhance the hydrophilicity of the substrate and membrane [43]. The pictures of the porous substrates and prepared membranes are shown in Figure S1 (in the Supplementary Materials). The opaque porous substrates were shown to be changed into a transparent state after the pore-filling by in situ polymerization.

In this work, the cross-linking of the PFAEMs was finely controlled by varying the cross-linker (DVB) contents. Some important membrane parameters (i.e. IEC, ER, contact angle, and WU) were correlated with the cross-linker content and the results are depicted in Figure 4a–d. As the DVB content increased, the IEC and WU values were shown to decrease, while the EARs and surface contact angles increased, demonstrating the reduction of free volume and number of hydrophilic fixed charges in the membranes. The images of the surface contact angle measurements are also displayed in Figure S2 (in the Supplementary Materials). The IECs of the PFAEMs fabricated with different porous PTFE substrate films (i.e., hydrophobic and hydrophilic grades) were almost the same at the identical membrane composition, as shown in Figure 4a. This result could prove that the pore size and porosity of the substrate films used, are comparable with each other. However, the PFAEMs prepared using a hydrophilic PTFE substrate (i.e., hydrophilic-PFAEMs) showed much lower EARs compared with those of the hydrophobic PTFE-based PFAEMs (i.e., hydrophobic-PFAEMs), meaning more facilitated ion transport through the more hydrophilic medium. The EAR values of the hydrophilic-PFAEMs started to sharply increase when adding the cross-linker of above 10 wt%, as shown in Figure 4b. Therefore, the optimal cross-linker content was determined as 10 wt%.

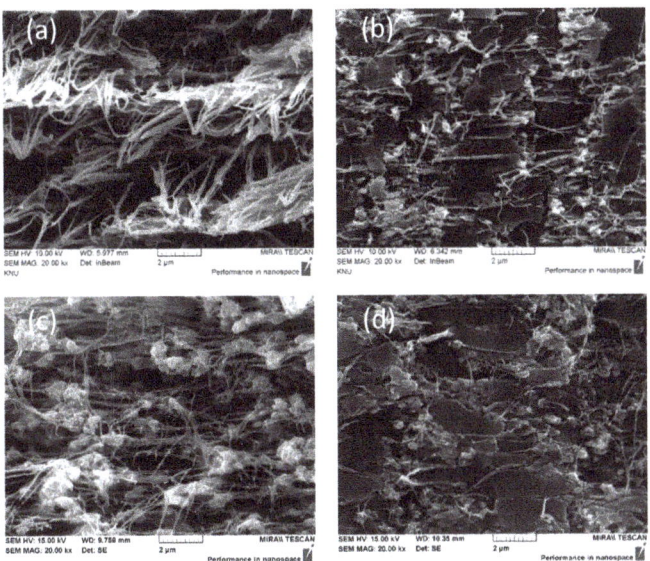

Figure 3. Cross-sectional FE-SEM images of porous substrates: ((**a**) hydrophobic PTFE; (**c**) hydrophilic PTFE)) and pore-filled anion-exchange membranes ((**b**) hydrophobic-PFAEM; (**d**) hydrophilic-PFAEM).

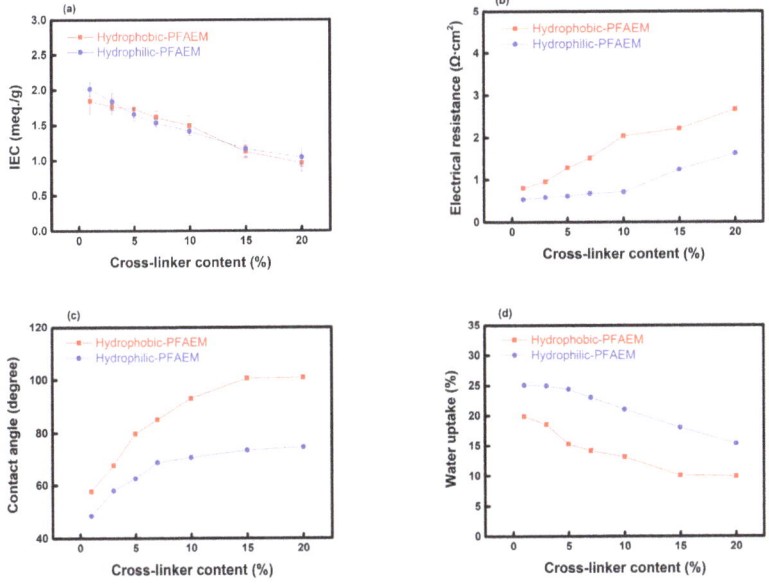

Figure 4. Changes in (**a**) IEC values, (**b**) electrical resistances (measured in a 0.5 M NaCl), (**c**) surface contact angles, and (**d**) water uptake of the PFAEMs by varying the content of cross-linking agent, divinylbenzene (DVB).

The various properties of the PFAEMs, which were fabricated with different porous substrate films and identical monomer composition (10 wt% DVB), are compared with those of the commercial membrane (AMX) in Table 2. Note that the same PFAEMs have also been utilized for comparative studies in ADLFC and VRFB systems. The surface contact angles of the PTFE based PFAEMs are shown to be much higher than that of the commercial membrane. Meanwhile, the contact angle of the hydrophilic-PFAEM is much smaller than that of the hydrophobic-PFAEM. This is one of the intrinsic characteristics of pore-filling types of membranes employing an inert porous substrate, that is, the hydrophobic nature of the porous substrate dominates the surface contact angles. The IEC values of the compared membranes were almost the same as each other, while the hydrophilic PFAEM revealed much higher conductivity for hydroxyl ions than those of both the hydrophobic-PFAEM and commercial membrane. As a result, the EAR of the hydrophilic-PFAEM was shown to be reduced by about four times compared with that of both the hydrophobic-PFAEM and the commercial membrane, because of the relatively high conductivity and thin membrane thickness. The prepared membranes exhibited excellent transport numbers for an anion (Cl^-), which were superior to that of the commercial membrane. Moreover, the alkaline stability of the AEMs was also checked via soaking tests under a harsh alkaline condition (i.e., 1 M KOH/60 °C/500 h). The transport numbers of the AEMs were shown to be significantly reduced after the alkaline soaking tests. The decrement in the transport numbers could have mainly originated from the degradation of quaternary ammonium sites under a harsh alkaline environment. The decrease in the transport number of the PTFE-based PFAEMs was much smaller than that of the commercial membrane, demonstrating that both, the use of chemically stable PTFE substrates, and the highly cross-linked ionomer, could largely enhance the alkaline stability of the membranes. The oxidative stability of the commercial and prepared membranes was also evaluated with the soaking experiment, using Fenton's reagent (3% H_2O_2 containing 2 ppm $FeSO_4$). As shown in Table 2 and Figure 5, the chemical stabilities of the PFAEMs were superior to that of the commercial membrane. The result demonstrates that the combination of a chemically stable PTFE substrate and a highly cross-linked hydrocarbon ionomer significantly enhances the chemical stability of the membranes. In addition, the differences in the chemical stability of the two PFAEMs were not that significant.

Table 2. Various characteristics of commercial and prepared membranes.

Membranes	Thickness (μm)	Contact Angle (Degree)	WU (%)	IEC (meq./g)	σ [1] (S/cm)	EAR [2] ($\Omega\ cm^2$)	t_- [3] (-)	t_- [4] (-)	WL [5] (%)
AMX (Astom Corp.)	135	44.8	21.1	1.40	0.015	0.93	0.975	0.853	10.2
Hydrophobic-PFAEM	82	93.2	13.2	1.46	0.010	0.84	0.984	0.926	0.32
Hydrophilic-PFAEM	40	70.7	21.1	1.42	0.019	0.21	0.990	0.954	0.45

[1] Membrane conductivity obtained by 2-point probe impedance measurement (in a 1.0 M KOH aqueous solution at 25 °C). [2] Membrane electrical resistance measured using a clip cell and an impedance analyzer (in a 1.0 M KOH aqueous solution at 25 °C). [3] Transport number for anion (Cl^-) determined by emf method (in 0.001/0.005 M NaCl aqueous solutions) (initially measured). [4] Transport number for anion (Cl^-) determined by emf method (in 0.001/0.005 M NaCl aqueous solutions) (measured after 500 h in the alkaline stability test). [5] Weight loss (%) after the soaking test using the Fenton's reagent at 80 °C for 1 h.

The MEAs utilizing two different PFAEMs were evaluated by the I–V polarization test, using a liquid fuel of 4 M KOH and 4 M N_2H_4 at 60 °C and 100% RH. The I–V and current–power (I–P) polarization curves of the MEAs are displayed in Figure 6. As a result, the power generation performance of the MEA was dramatically improved by employing the hydrophilic membrane instead of the hydrophobic one. The maximum power density of the MEA employing the hydrophilic-PFAEM was shown to be about 400 mW cm^{-2} at 1 A cm^{-2}. This result is almost comparable with that of the state-of-the-art membranes such as the Tokuyama A901 [28]. Since the crossover of liquid fuel through a membrane largely affects the energy conversion efficiency in such types of fuel cell [44], we also evaluated the diffusion coefficients of N_2H_4 through the PFAEMs, by means of conventional two-compartment diffusion cell tests. As shown in Table 3, the diffusion coefficient of the fuel molecule through the hydrophilic-PFAEM was revealed to be somewhat higher than that of the hydrophobic-PFAEM. It is believed that the more hydrophilic nature of the membrane increases the diffusion rate of the hydrophilic molecules. This means that the energy conversion efficiency of the hydrophilic-PFAEM is expected to be poorer than that of the hydrophobic-PFAEM, in terms of the fuel crossover rate. Therefore, it could be concluded that the dramatic improvement of the power density by employing the hydrophilic-PFAEM mainly resulted from the facilitated hydroxyl ion transport through the membrane.

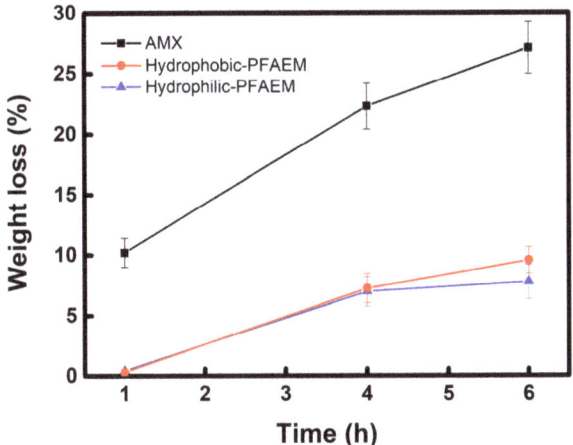

Figure 5. Time course changes in weight loss during the Fenton oxidation tests of anion-exchange membranes (AEMs).

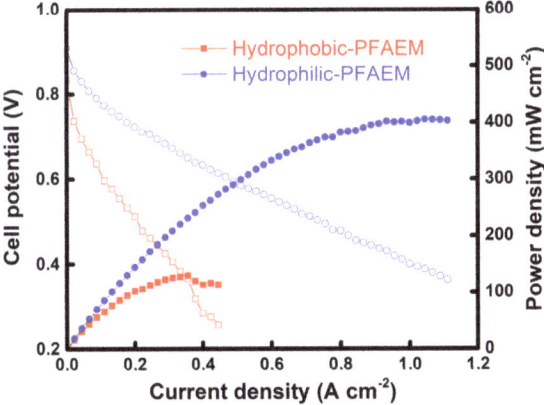

Figure 6. Power generation performances of the MEAs employing different PFAEMs (open symbols = cell potential; closed symbols = power density).

Table 3. Diffusion coefficients of hydrazine (N_2H_4) through the PFAEMs prepared by using different porous PTFE substrates.

Membranes	Diffusion Coefficient ($\times 10^9$, $cm^2\ s^{-1}$)
Hydrophobic-PFAEM	5.75
Hydrophilic- PFAEM	6.99

VRFB experiments were also performed to investigate the influence of the membranes on the battery characteristics, as shown in Figure 7. The charge—discharge performances were revealed to be largely affected by the membrane properties, and the efficiencies are summarized in Table 4. The hydrophobic-PFAEM showed the highest value of coulombic efficiency (CE) among the membranes tested, indicating that the crossover of redox species through the membrane was effectively suppressed owing to its high cross-linking degree and hydrophobic nature. The crossover rate of vanadyl ions (VO^{2+}) through the membranes could be estimated from a two-compartment diffusion cell experiment, by recording the time-course change of VO^{2+} concentration (Figure 8). The overall dialysis coefficient (K_A) values calculated from Equation (7) are also summarized in Table 4. The hydrophobic-PFAEM showed almost similar K_A values, despite its considerably reduced thickness compared to that of the commercial AMX membrane. However, the hydrophilic-PFAEM revealed a K_A value increased by about three times, owing to its hydrophilic nature and much-reduced thickness compared to the hydrophobic-PFAEM. As a result, it can be seen that the lowest CE value of the hydrophilic-PFAEM among the membranes compared, is due to the high crossover rate of the vanadium redox species. On the other hand, the hydrophilic-PFAEM exhibited the highest voltage efficiency (VE) among the membranes tested, due to the lowest mass transport resistance (note the data in Table 2). Overall, the VRFB employing the hydrophobic-PFAEM showed the highest energy efficiency (EE), of about 87%. In the case of ADLFC, it was preferable to use a hydrophilic-membrane because ion conductivity was the most critical factor determining the efficiency of the system. However, unlike ADLFC, the crossover of the redox species through the membrane more significantly influenced the system efficiency in VRFB. Therefore, in this case, it was found that the employment of a hydrophobic-PTFE-based PFAEM can result in a more improved energy efficiency.

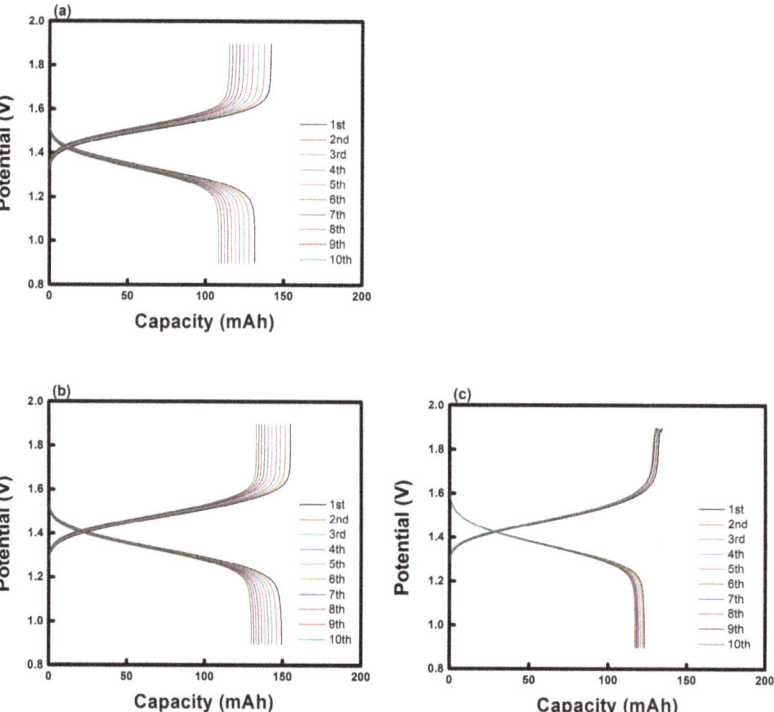

Figure 7. VRFB charge–discharge performances of (**a**) AMX, (**b**) Hydrophobic-PTFE based PFAEM, and (**c**) Hydrophilic-PTFE based PFAEM.

Table 4. Various characteristics of commercial and prepared membranes.

Membranes	CE (%)	VE (%)	EE (%)	K_A (×10^6, m s^{-1})
AMX (Astom Corp.)	93.6	87.7	82.1	2.16
Hydrophobic-PFAEM	97.2	89.4	86.9	2.63
Hydrophilic-PFAEM	90.9	91.3	83.0	7.69

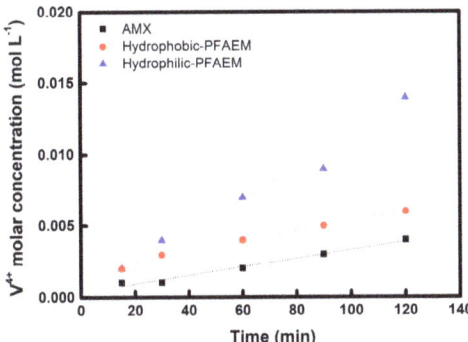

Figure 8. Time course change of vanadium ion transported through AMX, Hydrophobic-PTFE based PFAEM, and Hydrophilic-PTFE based PFAEM, respectively.

The soaking tests for the membranes were also conducted in 0.1 M V_2O_5 solution (in 5 M H_2SO_4) at 40 °C for about 200 h, to evaluate the oxidative stability in a vanadium electrolyte solution, as displayed in Figure 9. The results show that the V^{4+} ion concentration increased continuously owing to the oxidative degradation of the polymer. However, the PTFE-based PFAEMs exhibited a much-reduced increase rate of V^{4+} ion concentration, compared to that of the commercial AMX membrane. In addition, it was shown that the difference in hydrophilic properties of the porous substrates did not appear to have a significant effect in this case. These are well coincident with the results of the Fenton test and demonstrate that the PTFE-based PFAEMs have excellent oxidative stabilities in the harsh conditions of both ADLFC and VRFB. The excellent oxidative stability for the PFAEMs could be attributed to the use of chemically stable PTFE substrates and the decreased free volume due to the high cross-linking density, which reduces the influence of oxygen radicals on the polymer degradation [45].

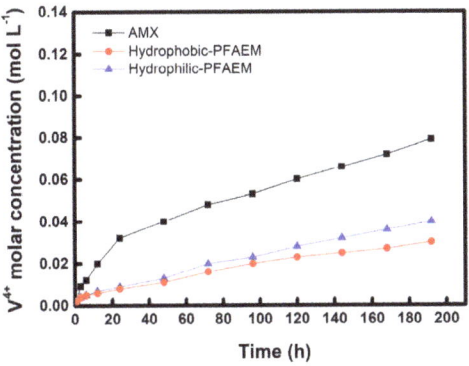

Figure 9. Time course changes in V^{4+} ion concentration during the oxidative stability tests of AEMs.

4. Conclusions

In this work, high-performance PFAEMs were successfully developed by combining a thin porous PTFE substrate and anion-exchangeable polymers with a double cross-linking structure, for application to electrochemical energy conversion systems, such as ADLFC and VRFB. In particular, two different kinds of porous PTFE substrates (i.e., hydrophilic and hydrophobic grades) were utilized for the fabrication of the PFAEMs. The PFAEMs exhibited excellent electrochemical characteristics and chemical stabilities, both in strong alkaline and oxidative conditions. In addition, the optimum membrane composition was investigated by adjusting the cross-linking degree. From the correlation studies, the membrane characteristics were systematically analyzed, and as a result, the optimal cross-linker (DVB) content was determined as 10 wt%. It was also proven that the use of hydrophilic PTFE porous substrate (rather than hydrophobic grade) can significantly enhance the power generation of ADLFCs, mainly due to the greatly facilitated hydroxyl ion transport through the membrane. As a result, an excellent maximum power density of around 400 mW cm^{-2} at 1 A cm^{-2}, which is almost comparable with that of the state-of-the-art membrane, was achieved by employing the hydrophilic PTFE-based PFAEM. The PFAEMs were also applied to VRFB for electrochemical energy storage. The results revealed that the crossover of vanadium redox species through the membrane most significantly affects the system efficiency in VRFB. The VRFB employing the hydrophobic-PFAEM exhibited the highest energy efficiency (EE), of 87%, among the membranes tested (cf. hydrophilic-PFAEM = 83% and AMX = 82%), mainly owing to its low crossover rate for vanadium redox ions and moderate membrane resistance. The PTFE-based PFAEMs also showed excellent oxidative stabilities in a highly acidic vanadium solution, which were superior to that of the commercial AMX membrane. Consequently, through this study, high-performance AEMs capable of long-term use under harsh alkaline and acidic conditions have been developed through the combination of porous PTFE substrates and an ionomer having both a high cross-linking degree and IEC. In particular, it was also revealed that the characteristics (e.g., hydrophilicity) of the porous substrate are critical and should match the operating environment for successful applications to electrochemical energy conversion processes.

Supplementary Materials: The following are available online at http://www.mdpi.com/1996-1073/13/18/4761/s1. Figure S1. Pictures of porous substrates (a) hydrophobic polytetrafluoroethylene (PTFE); (b) hydrophilic PTFE and pore-filled anion-exchange membranes (c) hydrophobic-pore-filled anion-exchange membranes (PFAEM); (d) hydrophilic-PFAEM. Figure S2. Surface contact angles of pore-filled anion-exchange membranes.

Author Contributions: Conceptualization, M.-S.K.; methodology, M.-S.K.; experimentation, D.-H.K. and M.-S.K.; validation, M.-S.K.; investigation, D.-H.K. and M.-S.K.; resources, M.-S.K.; writing-original draft preparation, D.-H.K. and M.-S.K.; writing—review and editing, M.-S.K.; supervision, M.-S.K.; project administration, M.-S.K.; funding acquisition, M.-S.K. All authors have read and agreed to the published version of the manuscript.

Funding: This research was funded by a 2019 Research Grant from Sangmyung University.

Acknowledgments: This research was supported by a 2019 Research Grant from Sangmyung University.

Conflicts of Interest: The authors declare no conflict of interest.

References

1. Tanaka, Y. Ion-exchange membrane electrodialysis for saline water desalination and its application to seawater concentration. *Ind. Eng. Chem. Res.* **2011**, *50*, 7494–7503. [CrossRef]
2. Parsa, N.; Moheb, A.; Mehrabani-Zeinabad, A.; Masigol, M.A. Recovery of lithium ions from sodium-contaminated lithium bromide solution by using electrodialysis process. *Chem. Eng. Res. Des.* **2015**, *98*, 81–88. [CrossRef]
3. Benvenuti, T.; Krapf, R.S.; Rodrigues, M.A.S.; Bernardes, A.M.; Zoppas-Ferreira, J. Recovery of nickel and water from nickel electroplating wastewater by electrodialysis. *Sep. Purif. Technol.* **2014**, *129*, 106–112. [CrossRef]
4. Mao, F.; Zhang, G.; Tong, J.; Xu, T.; Wu, Y. Anion exchange membranes used in diffusion dialysis for acid recovery from erosive and organic solutions. *Sep. Purif. Technol.* **2014**, *122*, 376–383. [CrossRef]
5. Kim, D.-H.; Park, J.-H.; Seo, S.-J.; Park, J.-S.; Jung, S.; Kang, Y.S.; Choi, J.-H.; Kang, M.-S. Development of thin anion-exchange pore-filled membranes for high diffusion dialysis performance. *J. Membr. Sci.* **2013**, *447*, 80–86. [CrossRef]
6. Palakkal, V.M.; Rubio, J.E.; Lin, Y.J.; Arges, C.G. Low-resistant ion-exchange membranes for energy efficient membrane capacitive deionization. *ACS Sustain. Chem. Eng.* **2018**, *6*, 13778–13786. [CrossRef]

7. Kim, D.-H.; Choi, Y.-E.; Park, J.-S.; Kang, M.-S. Capacitive deionization employing pore-filled cation-exchange membranes for energy-efficient removal of multivalent cations. *Electrochim. Acta* **2019**, *295*, 164–172. [CrossRef]
8. Legrand, L.; Shu, Q.; Tedesco, M.; Dykstra, J.E.; Hamelers, H.V.M. Role of ion exchange membranes and capacitive electrodes in membrane capacitive deionization (MCDI) for CO_2 capture. *J. Colloid Interface Sci.* **2020**, *564*, 478–490. [CrossRef]
9. Fang, J.; Lyu, M.; Wang, X.; Wu, Y.; Zhao, J. Synthesis and performance of novel anion exchange membranes based on imidazolium ionic liquids for alkaline fuel cell applications. *J. Power Sources* **2015**, *284*, 517–523. [CrossRef]
10. Kraytsberg, A.; Ein-Eli, Y. Review of advanced materials for proton exchange membrane fuel cells. *Energy Fuels* **2014**, *28*, 7303–7330. [CrossRef]
11. Kim, D.-H.; Park, J.-S.; Choun, M.; Lee, J.; Kang, M.-S. Pore-filled anion-exchange membranes for electrochemical energy conversion applications. *J. Electrochim. Acta* **2016**, *222*, 212–220. [CrossRef]
12. Zheng, Y.; Omasta, T.J.; Peng, X.; Wang, L.; Varoe, J.R.; Pivovar, B.S.; Mustain, W.E. Quantifying and elucidating the effect of CO_2 on the thermodynamics, kinetics and charge transport of AEMFCs. *Energy Environ. Sci.* **2019**, *12*, 2806–2819. [CrossRef]
13. Huang, G.; Mandal, M.; Peng, X.; Yang-Neyerlin, A.C.; Pivovar, B.S.; Mustain, W.E.; Kohl, P.A. Composite poly(norbornene) anion conducting membranes for achieving durability, water management and high power (3.4 W/cm^2) in hydrogen/oxygen alkaline fuel cells. *J. Electrochem. Soc.* **2019**, *166*, F637–F644. [CrossRef]
14. Lee, M.-S.; Kim, H.-K.; Kim, C.-S.; Suh, H.-Y.; Nahm, K.-S.; Choi, Y.-W. Thin pore-filled ion exchange membranes for high power density in reverse electrodialysis: Effects of structure on resistance, stability, and ion selectivity. *ChemistrySelect* **2017**, *2*, 1974–1978. [CrossRef]
15. Hong, J.G.; Zhang, B.; Glabman, S.; Uzal, N.; Dou, X.; Zhang, H.; Wei, X.; Chen, Y. Potential ion exchange membranes and system performance in reverse electrodialysis for power generation: A review. *J. Membr. Sci.* **2015**, *486*, 71–88. [CrossRef]
16. Zhang, B.; Zhang, E.; Wang, G.; Yu, P.; Zhao, Q.; Yao, F. Poly(phenyl sulfone) anion exchange membranes with pyridinium groups for vanadium redox flow battery applications. *J. Power Sources* **2015**, *282*, 328–334. [CrossRef]
17. Kim, D.-H.; Seo, S.-J.; Lee, M.-J.; Park, J.-S.; Moon, S.-H.; Kang, Y.S.; Choi, Y.-W.; Kang, M.-S. Pore-filled anion-exchange membranes for non-aqueous redox flow batteries with dual-metal-complex redox shuttles. *J. Membr. Sci.* **2014**, *454*, 44–50. [CrossRef]
18. Song, H.-B.; Kim, D.-H.; Kang, M.-S. Thin reinforced poly(2,6-dimethyl-1,4-phenylene oxide)-based anion-exchange membranes with high mechanical and chemical stabilities. *Chem. Lett.* **2019**, *48*, 1500–1503. [CrossRef]
19. Wang, Y.; Wang, S.; Xiao, M.; Song, S.; Han, D.; Hickner, M.A.; Meng, Y. Amphoteric ion exchange membrane synthesized by direct polymerization for vanadium redox flow battery application. *Int. J. Hydrog. Energy* **2014**, *39*, 16123–16131. [CrossRef]
20. Peighambardoust, S.J.; Rowshanzamir, S.; Amjadi, M. Review of the proton exchange membranes for fuel cell applications. *Int. J. Hydrog. Energy* **2010**, *35*, 9349–9384. [CrossRef]
21. Kirubakaran, A.; Jain, S.; Nema, R.K. A review on fuel cell technologies and power electronic interface. *Renew. Sust. Energy Rev.* **2009**, *13*, 2430–2440. [CrossRef]
22. Hwang, H.; Hong, S.; Kim, D.-H.; Kang, M.-S.; Park, J.-S.; Uhm, S.; Lee, J. Optimistic performance of carbon-free hydrazine fuel cells based on controlled electrode structure and water management. *J. Energy Chem.* **2020**, *51*, 175–181. [CrossRef]
23. Brouzgou, A.; Podias, A.; Tsiakaras, P. PEMFCs and AEMFCs directly fed with ethanol: A current status comparative review. *J. Appl. Electrochem.* **2013**, *43*, 119–136. [CrossRef]
24. Benipal, N.; Qi, J.; Gentile, J.C.; Li, W. Direct glycerol fuel cell with polytetrafluoroethylene (PTFE) thin film separator. *Renew. Energy* **2017**, *105*, 647–655. [CrossRef]
25. Bartrom, A.M.; Haan, J.L. The direct formate fuel cell with an alkaline anion exchange membrane. *J. Power Sources* **2012**, *214*, 68–74. [CrossRef]
26. Zeng, L.; Tang, Z.K.; Zhao, T.S. A high-performance alkaline exchange membrane direct formate fuel cell. *Appl. Energy* **2014**, *115*, 405–410. [CrossRef]

27. Merle, G.; Wessling, M.; Nijmeijer, K. Anion exchange membranes for alkaline fuel cells: A review. *J. Membr. Sci.* **2011**, *377*, 1–35. [CrossRef]
28. Pan, Z.F.; An, L.; Zhao, T.S.; Tang, Z.K. Advances and challenges in alkaline anion exchange membrane fuel cells. *Prog. Energy Combust.* **2018**, *66*, 141–175. [CrossRef]
29. Henkensmeier, D.; Najibah, M.; Harms, C.; Žitka, J.; Hnát, J.; Bouzek, K. Overview: State-of-the art commercial membranes for anion exchange membrane water electrolysis. *J. Electrochem. Energy Conv. Stor.* **2020**, *18*, 024001. [CrossRef]
30. Tsehaye, M.T.; Alloin, F.; Iojoiu, C. Prospects for anion-exchange membranes in alkali metal-air batteries. *Energies* **2019**, *12*, 4702. [CrossRef]
31. Ulaganathan, M.; Aravindan, V.; Yan, Q.; Madhavi, S.; Skyllas-Kazacos, M.; Lim, T.M. Recent advancements in all-vanadium redox flow batteries. *Adv. Mater. Interfaces* **2015**, *3*, 1500309. [CrossRef]
32. Xing, D.; Zhang, S.; Yin, C.; Zhang, B.; Jian, X. Effect of amination agent on the properties of quaternized poly(phthalazinone ether sulfone) anion exchange membrane for vanadium redox flow battery application. *J. Membr. Sci.* **2010**, *354*, 68–73. [CrossRef]
33. Zeng, L.; Zhao, T.S.; Wei, L.; Zeng, Y.K.; Zhang, Z.H. Highly stable pyridinium-functionalized cross-linked anion exchange membranes for all vanadium redox flow batteries. *J. Power Sources* **2016**, *331*, 452–461. [CrossRef]
34. Zhang, S.; Yin, C.; Xing, D.; Yang, D.; Jian, X. Preparation of chloromethylated/quaternized poly(phthalazinone ether ketone) anion exchange membrane materials for vanadium redox flow battery applications. *J. Membr. Sci.* **2010**, *363*, 243–249. [CrossRef]
35. Kim, J.-H.; Ryu, S.; Maurya, S.; Lee, J.-Y.; Sung, K.-W.; Lee, J.-S.; Moon, S.-H. Fabrication of a composite anion exchange membrane with aligned ion channels for a high-performance non-aqueous vanadium redox flow battery. *RSC Adv.* **2020**, *10*, 5010–5025. [CrossRef]
36. Yamaguchi, T.; Miyata, F.; Nakao, S. Pore-filling type polymer electrolyte membranes for a direct methanol fuel cell. *J. Membr. Sci.* **2003**, *214*, 283–292. [CrossRef]
37. Hwang, D.S.; Sherazia, T.A.; Sohn, J.Y.; Noh, Y.C.; Park, C.H.; Guiver, M.D.; Lee, Y.M. Anisotropic radio-chemically pore-filled anion exchange membranes for solid alkaline fuel cell (SAFC). *J. Membr. Sci.* **2015**, *495*, 206–215. [CrossRef]
38. Zhao, Y.; Yu, H.; Xie, F.; Liu, Y.; Shao, Z.; Yi, B. High durability and hydroxide ion conducting pore-filled anion exchange membranes for alkaline fuel cell applications. *J. Power Sources* **2014**, *269*, 1–6. [CrossRef]
39. Yang, S.; Choi, Y.-W.; Choi, J.; Jeong, N.; Kim, H.; Nam, J.-Y.; Jeong, H. R2R Fabrication of pore-filling cation-exchange membranes via one-time impregnation and their application in reverse electrodialysis. *ACS Sustain. Chem. Eng.* **2019**, *7*, 12200–12213. [CrossRef]
40. Palatý, Z.; Bendová, H. Numerical error analysis of mass transfer measurements in batch dialyzer. *Desalin. Water Treat.* **2011**, *26*, 215–225. [CrossRef]
41. Kim, S.; Tighe, T.B.; Schwenzer, B.; Yan, J.; Zhang, J.; Liu, J.; Yang, Z.; Hickner, M.A. Chemical and mechanical degradation of sulfonated poly(sulfone) membranes in vanadium redox flow batteries. *J. Appl. Electrochem.* **2011**, *41*, 1201–1213. [CrossRef]
42. Hu, G.; Wang, Y.; Ma, J.; Qiu, J.; Peng, J.; Li, J.; Zhai, M. A novel amphoteric ion exchange membrane synthesized by radiation-induced grafting α-methylstyrene and N,N-dimethylaminoethyl methacrylate for vanadium redox flow battery application. *J. Membr. Sci.* **2012**, *407–408*, 184–192. [CrossRef]
43. Specification Sheet for Hydrophilic PTFE Membrane Filters. Available online: http://advantecmfs.com/filtration/membranes/mb_ptfephil.php (accessed on 11 September 2020).
44. Gopi, K.H.; Bhat, S.D. Anion exchange membrane from polyvinyl alcohol functionalized with quaternary ammonium groups via alkyl spacers. *Ionics* **2018**, *24*, 1097–1109. [CrossRef]
45. Hu, B.; Miao, L.; Bai, Y.; Lü, C. Facile construction of crosslinked anion exchange membranes based on fluorenyl-containing polysulfone via click chemistry. *Polym. Chem.* **2017**, *8*, 4403–4413. [CrossRef]

© 2020 by the authors. Licensee MDPI, Basel, Switzerland. This article is an open access article distributed under the terms and conditions of the Creative Commons Attribution (CC BY) license (http://creativecommons.org/licenses/by/4.0/).

Article

Optimization of Perfluoropolyether-Based Gas Diffusion Media Preparation for PEM Fuel Cells

Riccardo Balzarotti [1], Saverio Latorrata [2,*], Marco Mariani [3], Paola Gallo Stampino [2] and Giovanni Dotelli [2]

1. Department of Energy, Politecnico di Milano, Via Lambruschini 4, 20156 Milano, Italy; riccardo.balzarotti@polimi.it
2. Department of Chemistry, Materials and Chemical Engineering "Giulio Natta", Politecnico di Milano, Piazza Leonardo da Vinci 32, 20133 Milano, Italy; paola.gallo@polimi.it (P.G.S.); giovanni.dotelli@polimi.it (G.D.)
3. Department of Mechanical Engineering, Politecnico di Milano, Via La Masa 1, 20156 Milano, Italy; marco.mariani@polimi.it
* Correspondence: saverio.latorrata@polimi.it; Tel.: +39-02-2399-3190

Received: 12 March 2020; Accepted: 8 April 2020; Published: 10 April 2020

Abstract: A hydrophobic perfluoropolyether (PFPE)-based polymer, namely Fluorolink® P56, was studied instead of the commonly used polytetrafluoroethylene (PTFE), in order to enhance gas diffusion media (GDM) water management behavior, on the basis of a previous work in which such polymers had already proved to be superior. In particular, an attempt to optimize the GDM production procedure and to improve the microporous layer (MPL) adhesion to the substrate was carried out. Materials properties have been correlated with production routes by means of both physical characterization and electrochemical tests. The latter were performed in a single PEM fuel cell, at different relative humidity (namely 80% on anode side and 60%/100% on cathode side) and temperature (60 °C and 80 °C) conditions. Additionally, electrochemical impedance spectroscopy measurements were performed in order to assess MPLs properties and to determine the influence of production procedure on cell electrochemical parameters. The durability of the best performing sample was also evaluated and compared to a previously developed benchmark. It was found that a final dipping step into PFPE-based dispersion, following MPL deposition, seems to improve the adhesion of the MPL to the macro-porous substrate and to reduce diffusive limitations during fuel cell operation.

Keywords: PEMFC; MPL production; hydrophobic coatings; perfluoropolyether; gas diffusion layer; durability

1. Introduction

The continuous increase in world energy demand as well as in energy use per person is leading to new challenges in energy production. A sustainable approach to energy production and storage is likely to be followed, as the actual use of fossil fuels has become insufficient for meeting both energy demands and environmental requirements in terms of greenhouse gas emissions [1]. Hydrogen is the most valuable candidate to fulfill these requirements, as it allows for clean and efficient energy storage and production [2,3]. From the energy conversion point of view, fuel cells are very promising devices, as they are able to produce electricity and heat from multiple sources, without drawbacks from the point of view of emissions [4].

Among others, proton exchange membrane fuel cells (PEMFCs) are considered very promising due to their zero-emission energy production process. In addition to this, PEMFCs are characterized by a high efficiency, low operating temperature, compactness and fast response to load change [5–7].

Among the PEMFCs components, the gas diffusion medium (GDM) plays a crucial role, especially in terms of water management and reactants diffusion to electrodes. The GDM is composed of two parts: a gas diffusion layer (GDL), which introduces macro-porosity properties that optimize reactants distribution from the flow field channels to the catalyst layer, and a micro-porous layer (MPL) that reduces liquid flooding and contact resistances [8,9]. The latter component is usually obtained by blade coating a thin carbon-based layer on the GDL surface and its addition to cell assembly was reported to improve cell performance [10].

In view of the position of the GDM in the cell assembly and of its role in the device operation, some features can be identified: the GDM should be permeable to reactants and products and, at the same time, it should display good properties in terms of electrical and thermal conductivity [10–12]. Finally, good mechanical properties would be highly desirable, in order to prevent damages and a consequent performance drop. Concerning thermal and electrical conductivity, good performances are easily achieved by using carbon-based materials. On the contrary, permeability is a more trivial feature, as specific GDL properties are used, which sometimes compete with each other. Polymer content in GDLs is a typical example of this duality. In order to prevent mass transport limitations, GDLs should be permeable to reactants, allowing a proper supply of fuel and oxidants to the catalyst-coated membrane (CCM). In a similar way, cathode side flooding should be avoided by properly managing excess water removal. In order to achieve this, GDLs are usually made hydrophobic by adding stable fluorinated polymers [10]. In many cases, polytetrafluoroethylene (PTFE) is used in order to obtain the desired water-repellent behavior, with polymer-loading on the layer surface in a range between 5% and 30% by weight [13]; nonetheless, commercial components featuring up to 70% PTFE as both hydrophobic agent and binder have been employed in other valuable works [14,15]. Thus, the best performance as a function of polymer content is the result of two competing properties: if a small quantity of polymer is used, low values in terms of ohmic losses due to polymer dielectric properties will be present; nevertheless, at the same time, poor hydrophobic properties will be obtained.

GDLs are usually employed in the form of carbon paper or carbon cloths, with different properties in terms of reactants/products diffusion and pores size distribution [16–18]. The latter property is of particular interest to enhance device performance: an optimum in pores dimension should be found, as higher pores enhance species gas diffusion, but they increase ohmic losses [11]. In this view, the MPL is coated onto the GDL aiming to maximize the contact between the catalyst layer and the GDL. Additionally, transport properties, and thus electrochemical performance, are enhanced thanks to the introduction of a micro-porosity in the diffusion media [17]. In many works, the presence of the MPL in cell assembly was proven to be of remarkable importance for improving the overall device performance [8,10,17,19–22]; the hydrophobicity of the MPL material together with its micro-porosity was found to enhance the device performance, due to better properties in terms of water removal in the Membrane Electrode Assembly (MEA) [23–26].

The MPL is typically produced by mixing a carbon powder with a hydrophobic agent, solvents and surfactants in order to produce a carbon ink. Then, the carbon-based ink is deposited onto the GDL surface by using a slurry coating deposition technique (i.e., blade coating, tape casting or spray coating) and, finally, the GDM is heat treated to remove liquid components and to consolidate the MPL layer [10]. MPL adhesion on GDL is a crucial requirement to be fulfilled in order to reach the target of device durability. MPL detachment is a highly undesired phenomenon, which leads to a decrease in device performance due to catalyst deactivation and mass transfer limitations [9].

In this work, an anionic polyurethane polymer based on a perfluoropolyether (PFPE) backbone, which had been already proved to be a viable alternative to the commonly used PTFE [27], was employed as hydrophobic agent in GDMs production. Different manufacturing routes were investigated, aiming to maximize water management and adhesion properties. The achievement of these targets was coupled with the enhancement of the electrochemical performance of the device. Samples were characterized both from the physical and electrochemical point of view, in order to assess the feasibility of production processes.

2. Materials and Methods

2.1. Preparation

In order to obtain a water-repellent gas diffusion medium, both GDL and MPL have to be made hydrophobic. In the present work, such a goal was accomplished by using fluorinated polymers and by applying different procedures. In the case of GDLs, a macro-porous carbon cloth (SCCG 5N by SAATI Group, Appiano Gentile, Italy) was treated by means of a dipping/drying procedure in a fluorinated aqueous emulsion.

The MPL was produced by blade coating deposition of an ink precursor onto the GDL surface, according to a procedure reported in literature [28]. In a typical experiment, carbon black (Vulcan XC-72R, Cabot Italiana S.p.A., Ravenna, Italy) was dispersed into a solution containing distilled water and isopropyl alcohol (Sigma-Aldrich). After the addition of Fluorolink® P56 (Solvay Solexis, Milan, Italy), a perfluoropolyether (PFPE)-based polymer, as hydrophobic agent, the four components were mixed for 10 min at 8000 rpm, by using an UltraTurrax T25 homogenizer (IKA Instruments, Staufen, Germany). The employed concentration of the polymer with respect to the carbon black was 6 wt. %, which allowed the achievement of the highest performance in the reference [27].

Before applying MPLs coating, GDL substrates were pre-treated by dipping in a PFPE emulsion (1 wt. %), and different drying processes, depending on the specific preparation route, were carried out. Inks were deposited onto such pre-treated GDLs by using the blade coating technique. A K-Control Coater device (RK Print-Coat Instruments Ltd., Litlington, UK) was used; the gap between blade and substrate, corresponding to the wet thickness of the MPL, and the blade speed were set at 40 μm and 0.154 m s^{-1}, respectively. Different production paths were applied in order to optimize the MPL adhesion to the GDL substrate. A graphical representation of them is reported in Figure 1, while Table 1 shows the polymer amount employed for both GDLs and MPLs preparation; for the sake of comparison data about the reference sample [27] is also reported.

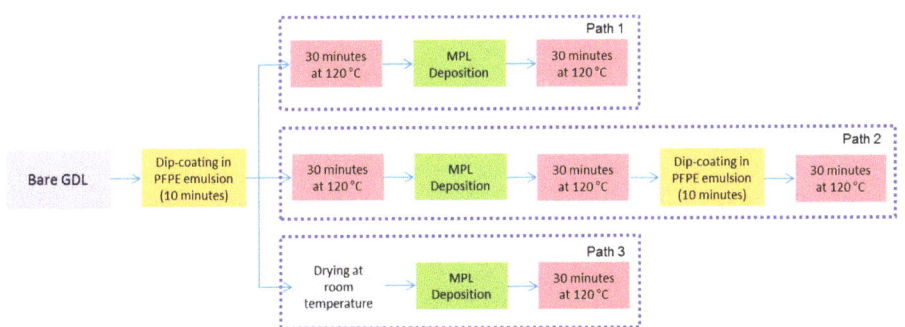

Figure 1. Schematic representation of GDMs production processes (same color corresponds to the same unit operation/step).

Table 1. Resume of the prepared samples: details on production path, hydrophobic agent and polymer content in both GDL and MPL.

Sample	GDL		MPL		Production Process
	Hydrophobic Agent	Content [wt. %]	Hydrophobic Agent	Content [wt. %]	
GDM6-p1	PFPE	1	PFPE	6	Path 1
GDM6-p2	PFPE	1	PFPE	6	Path 2
GDM6-p3	PFPE	1	PFPE	6	Path 3
GDM6-ref. [1]	PTFE	12	PFPE	6	reference [1]

[1] reference [27].

In Path 1, a common approach was used: in the first step, the GDL was made hydrophobic by dip coating in a 1 wt. % PFPE emulsion, then it was heat treated at 120 °C for 10 min. Such a temperature, that is much lower than the one employed in the heat treatment of common PTFE-containing samples (around 350 °C) [10], was selected on the basis of the authors' previous works [27,28], and was intended to eliminate solvent and surfactant only; indeed, it was found that treating amorphous PFPE-based polymers at a higher temperature would be meaningless [28]. Then, the MPL was deposited and the same thermal treatment at 120 °C was performed. In Path 2, following the same operation as for Path 1, a final dipping into the polymer emulsion followed by heat treatment was added, in order to enhance the MPL adhesion to the substrate. Path 3 represents a new approach to GDMs production as an attempt to maximize MPL-GDL interactions was made by performing a single heat treatment; indeed, the polymer-treated GDL was dried at room temperature and directly coated by the MPL; then, the obtained GDM was heat treated at 120 °C for 30 min.

2.2. Characterization

The surface hydrophobicity of GDMs was assessed by means of static contact angle measurements. For this purpose, an OCA 20 instrument (DataPhysics Instrument GmbH, Filderstadt, Germany) was used. Further details regarding contact angle measurements on PEMFCs components have been reported by the authors elsewhere [28].

The pores size distribution of the prepared samples was assessed through mercury intrusion porosimetry (MIP) by means of an Autopore V9600 by Micrometrics Instrument Corporation.

The produced samples were assembled in a single lab-scale fuel cell; in order to focus on GDMs properties only, a commercial catalyst coated membrane (CCM, provided by Baltic Fuel Cells GmbH, Germany) was employed for the electrochemical tests. Nafion 212 was the electrolytic membrane while platinum was the catalytic active phase, with different loadings at the anode side (0.2 mg cm^{-2}) and at the cathode side (0.4 mg cm^{-2}) due to different kinetics of oxidation and reduction reactions. The electrodes' active area was equal to 23 cm^2. Finally, graphitic bipolar plates were used for reagents distribution; a single feeding channel was used for supplying hydrogen, while a triple serpentine was used for air. Various operating conditions were tested, both in terms of temperature and humidity. Electrochemical experiments were carried out at 60 °C and 80 °C, with different inlet gas humidification levels in order to evaluate the water management properties of cell components. The relative humidity (RH) of the hydrogen fed to the anode was kept constant at 80%, while it was set at 60% and 100% for the cathodic air. Inlet volumetric flow rates were fixed at 0.25 Nl min^{-1} for hydrogen and 1 Nl min^{-1} for air; these values correspond to stoichiometric ratios of 1.3 and 2.2 for hydrogen and air, respectively, calculated at 1.2 A cm^{-2}.

Polarization curves were obtained by monitoring the fuel cell current, voltage and power during operation by using an electronic load (RBL488 50-150-800, TDI Power, Hackettstown, New Jersey, USA), in galvanostatic mode, from open circuit voltage (OCV) to high current density values with 0.09 A cm^{-2} steps. Electrochemical Impedance Spectroscopy (EIS) was used together with polarization curves in order to investigate the materials' electrochemical properties and behavior during cell operation. By means of a proper equivalent circuit model [29], EIS allows us to quantify the different contributions to cell potential losses; each of them is characteristic of physical-chemical processes which take place in the fuel cell, and they can be investigated to compare materials' properties. EIS was performed by means of a frequency response analyzer (FRA, Solartron 1260, Solartron Analytical, Farnborough, England, UK), which was connected to the electronic load. A typical experiment was performed in galvanostatic mode, in the frequency range 0.5 Hz-1 kHz [30]. The obtained spectra were fitted using the ZView software (Scribner Associates, Southern Pines, North Carolina, USA). Fuel cell internal losses were modeled using an equivalent electrical circuit which consisted of a resistance representing ohmic losses in series with two capacitance/resistance parallel circuits. The first parallel circuit was introduced in order to model charge transfer resistance by quantifying activation polarization, while the second one modelled mass transfer resistance by determining concentration polarization [31,32].

Due to the porous nature of the analyzed components, constant phase elements (CPE) were used as circuit elements instead of pure capacitances [33,34].

The durability of the best performing sample and of the PFPE-based benchmark [27] was evaluated. This was realized by operating the fuel cell at a constant current density (0.5 A cm^{-2}) for 1000 h at 60 °C and RH 80–100%, performing polarization tests every 168 h (i.e., one week).

The same samples were also subjected to an ex-situ mechanical accelerated stress test (AST) in order to have a faster and more real evaluation of the durability without carrying out tests for thousands of hours. As a matter of fact, mechanical degradation was proven to be the most critical stressor for GDMs, mainly due to detachment of the MPL surface carbon, which may be caused by both reactants flow and water [35]. The GDMs were assembled in a dummy cell with a 210 μm thick Teflon membrane as a separator without catalyst layers for preventing chemical stresses on the samples [12]. For the same reason, only air was supplied continuously to each side of the cell for 1000 h. Flow rates were 0.5 NL min^{-1} at the dummy anode and 2 NL min^{-1} at the dummy cathode, so twofold values compared to the ones used for standard electrochemical tests aiming to quicken mechanical degradation. Electrochemical tests in the running fuel cell were performed again upon the AST experiments, in order to evaluate the effects of the imposed stressors on the GDMs.

3. Results

3.1. Physical Characterization

Due to the strong influence of GDMs' water management properties on fuel cell performance, the components' hydrophobicity was evaluated using static contact angle analysis; the results of such measurements are reported in Table 2. Values were recorded both before cell testing (BCT) and after cell testing (ACT); in the latter case, measurements were performed at both the anode side and the cathode side. Ten measurements per sample were performed and then averaged.

All the samples were close to the superhydrophobicity limit (150°) upon preparation. While GDMs prepared by paths 2 and 3 show practically unchanged contact angle values upon electrochemical tests, both for anodic and cathodic samples, the GDM prepared by means of path 1 exhibited a dramatic decrease in hydrophobicity of the anodic sample. This behavior can be ascribed to the water back-diffusion taking place at a high current density due to unbalanced pressure and concentration between electrodic compartments; this sample was not able to withstand the unavoidable back-diffusion because of the possible low adhesion between MPL and GDL. Indeed, due to that faulty adhesion, part of the MPL surface carbon might have been lost and the measurement of the contact angle affected by the back layer which was treated with a lower quantity of PFPE. As a rough confirmation of this, the loss of material of the GDMs upon the electrochemical testing procedure was measured and reported in Table 2 as well. Such loss is mainly due to the detachment of the MPLs surface carbon. Of course, the cell disassembling procedure can increase the material loss, but this is true and always the same phenomenon for all the samples. GDM6-p2 and GDM6-ref. showed a probable better adhesion of the MPL to the GDL, since a loss that is much lower than the one of the other samples was observed.

Table 2. Measured values of static contact angle (C.A.) for GDM samples (BCT: before cell test, ACT: after cell test) and weight loss after cell tests.

Sample	C.A. BCT [°]	C.A. ACT [°]		Loss ACT [wt. %]
		Anode	Cathode	
GDM6-p1	152 ± 3	124 ± 9	152 ± 4	9.2
GDM6-p2	151 ± 2	149 ± 5	149 ± 3	0.9
GDM6-p3	150 ± 3	153 ± 3	153 ± 4	5.7
GDM6-ref. [27]	146 ± 5	154 ± 3	145 ± 2	1.0

The results of the porosimetry tests in terms of pores size distribution (Figure 2) show important differences between the samples. The porosity of the GDM is crucial because it influences the efficiency of gases and water transport across such components. The classification of GDMs pores does not correspond to the one adopted in other fields and is as follows: macropores (pores radius > 5 μm), mesopores (0.07 μm < radius < 5 μm), and micropores (radius < 0.07 μm) [17]. Accordingly, in Figure 2, it is possible to notice the presence of micropores which are more pronounced for GDM6-p2 than for the other new samples. Such behavior can result in a better water management, since a greater amount of micropores would enhance the capillary effect of the MPL, which could remove the excess water more quickly [17,36]; moreover, the average pore diameter of GDM6-p2 (around 40 nm) is lower than the one shown by the other new samples (in the range 47–50 nm), and very similar to that exhibited by GDM6-ref. Such a result will prove crucial for the electrochemical performance, especially in the high current density region, where a high amount of water is produced. Conversely, it must be also noticed that the macropores region is very similar for all the new samples, due to the presence of the same PFPE-based GDL substrate—independent of the process of dipping or thermal treatment followed. The difference with the reference sample in which the GDL had been treated with PTFE is clear, but the higher impact of macropores is on gas transport from the bipolar plates to the catalyst layer [12].

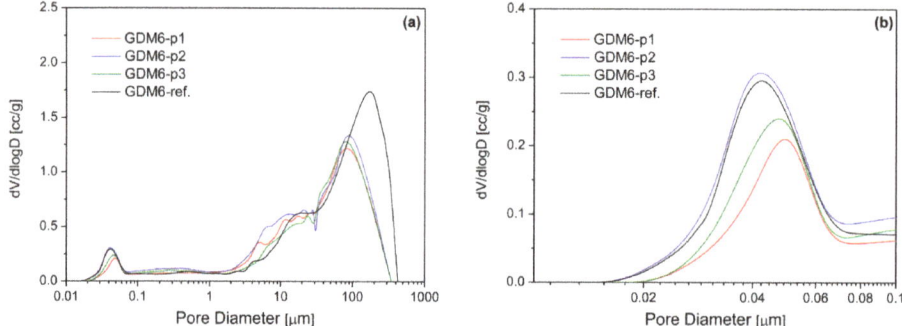

Figure 2. Pores size distribution of the prepared GDMs and the reference sample (**a**) and enlargement of the micro-porous region (**b**).

3.2. Electrochemical Characterization

Figure 3 shows the results of the electrical tests in terms of polarization and power density curves. It can be noticed that GDM6-p2 exhibited the best performance among the new prepared samples at all the operating conditions employed: both the highest output power density and the lowest slope of polarization curves in the whole range of generated current density were obtained with the fuel cell assembled with the samples. The curves related to GDM6-p2 performances are practically overlapped with the ones obtained with the benchmark; however, it is worth underlining that, in the reference sample, the GDL was PTFE-treated with a much higher polymer concentration (12 wt. % of PTFE vs. 1 wt. % of PFPE). In addition, the performances are stable on the whole range of conditions adopted for the testing, with just slight variations in terms of mass transfer resistance. This suggests that there is a proper management of the water content, independent of the operating conditions, and that the sample is not affected by significant degradation mechanisms during the short period.

The best performances of the GDM6-p3 sample were achieved at low relative humidity, with just a slight increase in ohmic losses, and mainly of losses due to concentration polarization compared to GDM6-p2. Overall, it is evident that a sharp improvement of performances for the fuel cell assembled with this sample occurs when the gas flow at the cathode features a relative humidity of 60%. This can be due to a mass transfer enhancement, considering that a reduced humidity prevents the water condensation within the cell, thus preserving the oxygen diffusivity in the cathodic compartment.

Finally, GDM6-p1 is the worst performing sample, in particular at high current density. Mass transfer losses are particularly large compared to the other samples and prevent the achievement of decent power densities, especially at 80 °C. Moreover, strong voltage drops have been recorded at medium current density, i.e., in the ohmic region, at 80 °C, which suggests a dependence of the electrolyte hydration on the operating temperature due to excessive vapor permeation through the MPL.

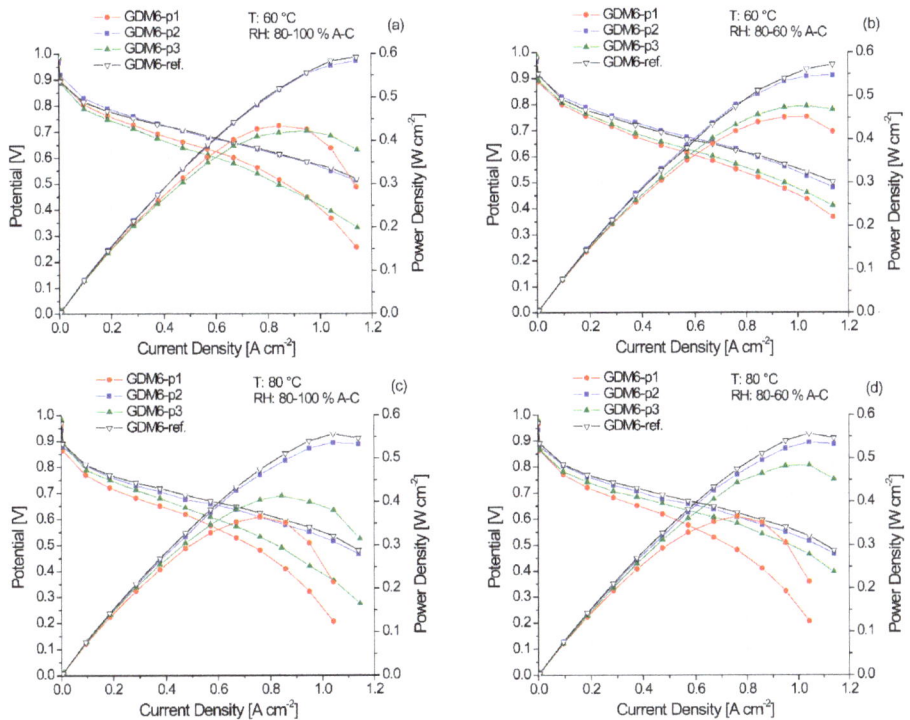

Figure 3. Polarization and power density curves obtained with fuel cells assembled with prepared PFPE-based GDMs. Operating conditions: 60 °C and RH (A–C) 80–100% (**a**), 60 °C and RH (A–C) 80–60% (**b**), 80 °C and RH (A–C) 80–100% (**c**), 80 °C and RH (A–C) 80–60% (**d**).

Figures 4 and 5 show trends of both ohmic and mass transfer resistance as a function of current density. Such parameters are those mostly influenced by GDMs features, whereas charge transfer resistance is mainly dependant on the catalytic layer, which is a commercial component with fixed properties in this work; therefore, it has not been reported.

The ohmic resistances shown in Figure 4 follow similar trends under all the operating conditions. GDM6-p2 exhibits the best behavior, comparable to the one of the benchmark. The difference with the other samples may be determined by the additional PFPE coating that could reduce the permeability of the MPL, thus enhancing the accumulation of water in the electrolyte at the cathodic side: this probably has favored the back-diffusion mechanism in the MEA, so the ohmic resistance of the ionomer has been kept low due to uniform and constant hydration. Indeed, at a low-medium current density, the water removal action of the MPL is of limited importance, given the low amount of water produced at the cathodic side; however, its presence is effective in preventing the dispersion of water expelled by the ionomer—particularly at higher temperatures and a low relative humidity. Such effects point out the dual role of the MPL, which at the same time is responsible for effective water removal in order to avoid the cell flooding and for maintaining a proper level of hydration for the electrolyte. The RH decrease in the gas flow at the cathode seems to be detrimental, mainly for the less efficient samples, while GDM6-p2 is definitely on par with the benchmark from this point of view.

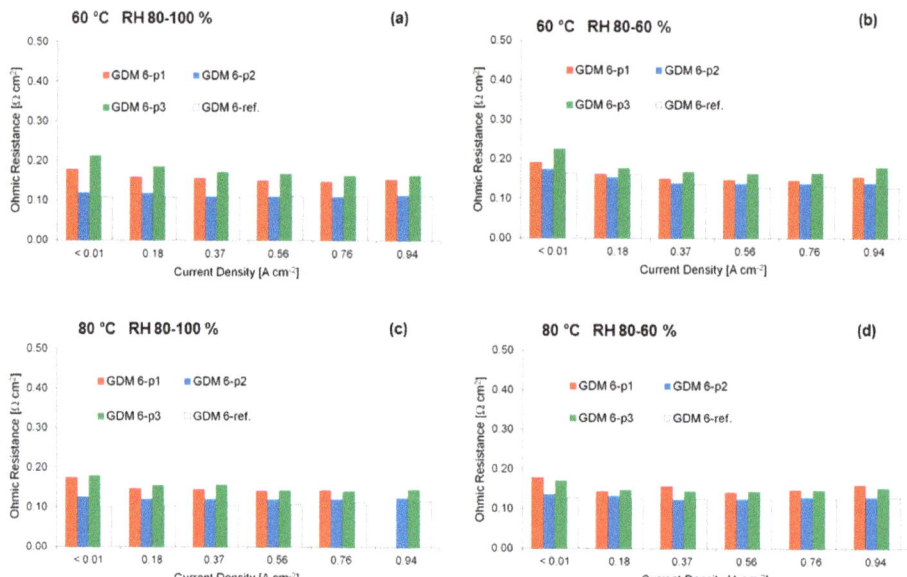

Figure 4. Trend of ohmic resistance as a function of current density obtained with fuel cells assembled with prepared PFPE-based GDMs. Operating conditions: 60 °C and RH (A-C) 80–100% (**a**), 60 °C and RH (A-C) 80–60% (**b**), 80 °C and RH (A-C) 80–100% (**c**), 80 °C and RH (A-C) 80–60% (**d**).

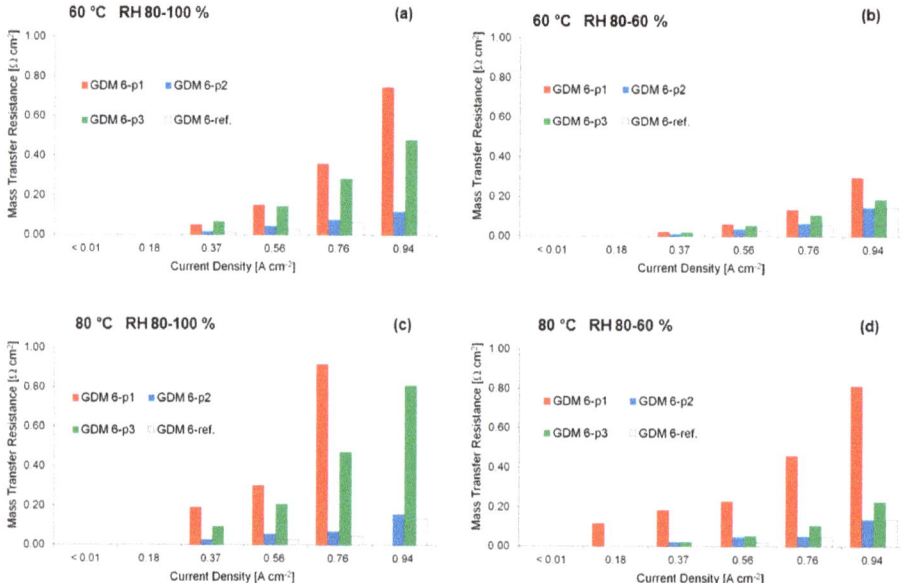

Figure 5. Trend of mass transfer resistance as a function of current density obtained with fuel cells assembled with prepared PFPE-based GDMs. Operating conditions: 60 °C and RH (A-C) 80–100% (**a**), 60 °C and RH (A-C) 80–60% (**b**), 80 °C and RH (A-C) 80–100% (**c**), 80 °C and RH (A-C) 80–60% (**d**).

Figure 5 shows that the most significant distinction between the samples are ascribed to the mass transfer resistance, which is deeply related to diffusion limitations arising mainly from water production within the cell. Indeed, it can be noticed that GDM6-p1 suffers from a sharp increase in resistance even at relatively low current density. This is in accordance with the ineffective water management induced by high diffusion limitations observed for polarization curves in Figure 3, and it could be due to the lower amount of PFPE on the MPL surface compared to GDM6-p2. Moreover, the preparation route might have hindered the adherence of the MPL to the substrate leading to the formation of water films at their interface, which acts as a barrier against oxygen transport and promotes the delamination of the GDM, as suggested by the higher material loss (Table 2). It is evident that the second preparation route is more beneficial in terms of mass transfer with respect to the third as well, maybe due to the addition of PFPE in the final step of the process which could induce a more effective adhesion between MPL and GDL.

Results of the preliminary durability tests performed with GDM6-p2 and the PFPE-based reference sample are reported in Figure 6 in terms of polarization and power density curves. Electrical tests were carried out every 168 h (i.e., one week) of running at constant current density, i.e., 0.5 A cm^{-2}. A good durability can be claimed since all the curves obtained for both samples are practically overlapped. However, a slight potential drop can be seen in the high current density region of the polarization curves obtained for the reference sample (Figure 6b). This may be traced back to a worsening of the water management capability, maybe due to the bigger macropores (Figure 2) of the GDL and to the fact that different polymers were in contact in GDL and MPL, therefore likely reducing the adhesion between the components. Indeed, the material loss upon disassembling the fuel cell, after performing the whole test, was 2.1 wt. % and 3.5 wt. % for GDM6-p2 and GDM6-ref., respectively. The bigger macropores may reduce the capillary condensation and water removal, since they can be more easily clogged by the produced water; indeed, such phenomena can be only observed at a high current density, when more water is being generated.

However, it is worth underlining that the PEMFC systems which are already commercialized produce electric energy in the ohmic region; in the region, both samples exhibited the same performance. This is also the reason why these tests were carried out at 0.5 A cm^{-2}. So, ad-hoc accelerated stress tests (AST) are needed in order to be more accurate in predicting the resistance against degradation of such components. Figure 7 shows the polarization and power density curves after 1000 h of AST compared to those obtained for fresh samples, i.e., not subjected to AST. Obviously, the stressed samples exhibited worse performance, even though the reduction was not dramatic and the voltage values in the ohmic zone were still capable of producing acceptable efficiencies in possible real systems [4,35]. As expected, the highest loss occurred in the concentration polarization region, i.e., at a high current density, and this is due to the difficult water management caused by the partial loss of the surface carbon of the MPL upon mechanical AST. Indeed, upon disassembling the fuel cell after performing ASTs, a total weight loss of 8.2 wt. % was found for the reference sample, while a smaller loss of 2.9 wt. % was detected for GDM6-p2. This may be a further indication of a satisfying adhesion between the new MPL and its GDL substrate, and of a better resistance to degradation compared to our benchmark featuring a PFPE-based MPL deposited onto a standard PTFE-treated GDL.

This may be more understandable by analyzing the trend of ohmic and, at a higher extent, of mass transfer resistances as a function of current density (Figure 8), obtained after performing AST. The change in the ohmic resistance (Figure 8a) upon AST is not significant for the new produced sample (GDM6-p2), while a sharper increase compared to the values of the fresh sample can be noticed for the reference sample (GDM6-ref.). This may have been caused by a worse contact between the MPL and the catalyst layer due to the witnessed loss of surface material, which would lead to an increase in the contact resistance and consequently in the overall ohmic resistance.

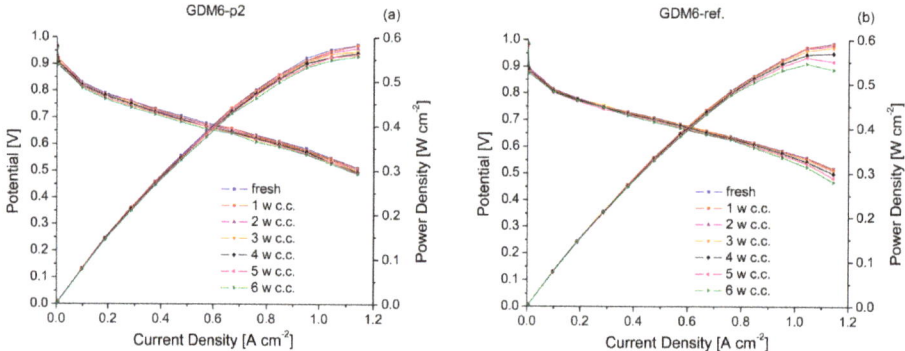

Figure 6. Polarization and power density curves obtained every week of constant current durability tests for GDM6-p2 (**a**) and GDM6-ref. (**b**) GDMs. Operating condition: 60 °C and RH (A-C) 80–100%.

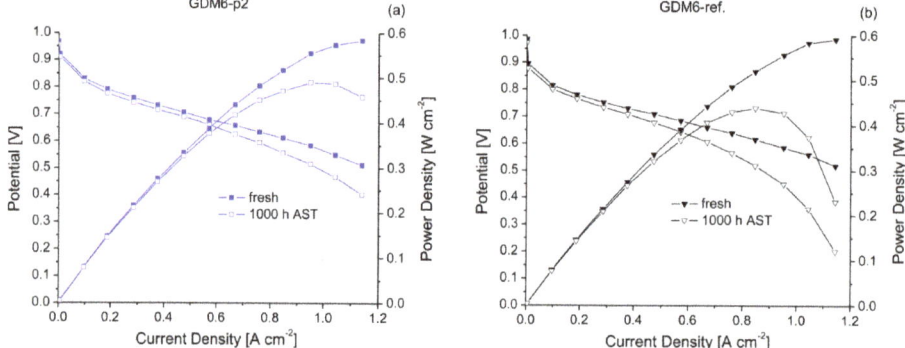

Figure 7. Polarization and power density curves obtained upon 1000 h of accelerated stress tests for GDM6-p2 (**a**) and GDM6-ref. (**b**) compared to curves obtained for fresh (as prepared) samples. Operating condition: 60 °C and RH (A-C) 80–100%.

On the other hand, an increase in mass transfer resistance (Figure 8b) is clear for both samples. This was largely expected, since mechanical degradation induced by AST has caused partial MPL material loss and worsened the capability of removing the excess water, as found in a previous work [35]. However, it is clear that the new sample was able to improve resistance to degradation and that it was damaged less than the reference sample; indeed, for GDM6-p2, a lower increase in the mass transfer resistance with respect to the fresh sample was found compared to the change in the same parameter for the benckmark, i.e., GDM6-ref. sample.

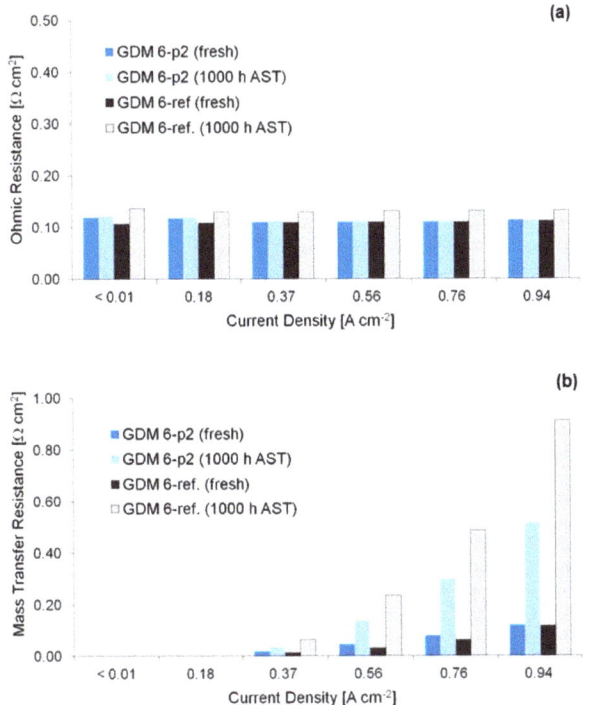

Figure 8. Trend of ohmic resistance (**a**) and mass transfer resistance (**b**) as a function of current density obtained for fresh samples (not subjectd to AST) and upon 1000 h of accelerated stress tests. Operating condition: 60 °C and RH (A-C) 80–100%.

4. Conclusions

In the present work, three different preparation routes were employed to prepare perfluoropolyether (PFPE)-based GDMs, with both GDLs and MPLs treated with the same polymer. Such polymers had been already applied successfully as alternative to the currently most-used hydrophobic agent, PTFE. The main improvement introduced by the novel polymer was the chance to operate at a much lower temperature during the preparation of the GDM, and to use a lower amount of hydrophobic agent compared to PTFE.

This work mainly aimed to investigate the electrochemical performance of fuel cells assembled with novel GDMs when PFPE was used for both GDLs and MPLs, while the benchmark was a sample in which only MPL was based on PFPE, keeping PTFE as hydrophobic agent for the GDL. Moreover, the reason why different preparation routes were carried out lies in the pursuit of an effective adhesion between MPL and GDL, with the final target of improving the durability of the GDM.

It was found that the preparation method had an effect on the pores size distribution, changing the microporous and macroporous region of the obtained sample. This was reflected in the electrochemical performance, too; indeed, much better polarization curves were achieved for GDM6-p2, which showed a better compromise between porosity, wettability and adhesion. Indeed, the sample—which featured a further dipping in the polymeric dispersion during the preparation, upon the MPL deposition using the blade coating technique—exhibited a low change in the static contact angle after electrochemical tests, together with a less pronounced loss of material upon the disassembly of the fuel cell.

The durability of the best performing sample and of the PFPE-based MPL benchmark was tested by applying both constant current tests and mechanical accelerated stress tests (ASTs) for 1000 h. Fuel cell tests performed after durability experiments revealed a better resistance of the novel sample compared to the reference one against degradation, since a reduced increase in mass transfer limitations—likely due to better adhesion between the MPL and the GDL—was found.

The use of PFPE is of great interest from an economic point of view too. The GDMs production procedure is comparable to that already employed for PTFE-based components, but the amount of material is much lower (1 wt.% vs. 12 wt.% for the GDLs treatment and 6 wt.% vs. 12 wt.% for the MPLs production). In addition, the maximum temperature of the heat treatment step is lower too, thus reducing the energy consumption and process time. The durability improvement is particularly meaningful, because it may extend the service life of the fuel cell, thus reducing the costs of maintenance and waste disposal.

These promising findings may prompt further detailed studies about the possible employment of PFPE as a valid alternative to PTFE for the hydrophobic treatment of both GDLs and MPLs, aiming to reduce the temperature of the thermal treatment in the preparation route, as well as to improve the durability of the final obtained GDMs.

Author Contributions: Individual contributions are as follows: conceptualization, P.G.S., S.L. and G.D.; methodology, S.L. and P.G.S.; investigation, R.B.; data curation, R.B., S.L. and M.M.; writing—original draft preparation, R.B. and S.L.; writing—review and editing, M.M. and S.L.; supervision, G.D. All authors have read and agreed to the published version of the manuscript.

Funding: This research received no external funding.

Conflicts of Interest: The authors declare no conflict of interest.

References

1. Omrani, R.; Shabani, B. Review of gas diffusion layer for proton exchange membrane-based technologies with a focus on unitised regenerative fuel cells. *Int. J. Hydrog. Energy* **2019**, *44*, 3834–3860. [CrossRef]
2. Ehteshami, S.M.M.; Chan, S.H. The role of hydrogen and fuel cells to store renewable energy in the future energy network—Potentials and challenges. *Energy Policy* **2014**, *73*, 103–109. [CrossRef]
3. Garland, N.L.; Papageorgopoulos, D.C.; Stanford, J.M. Hydrogen and fuel cell technology: Progress, challenges, and future directions. *Enrgy Proced* **2012**, *28*, 2–11. [CrossRef]
4. Barbir, F. *PEM Fuel Cells: Theory and Practice*, 2nd ed.; Academic Press: London, UK, 2013; p. 444.
5. Peighambardoust, S.J.; Rowshanzamir, S.; Amjadi, M. Review of the proton exchange membranes for fuel cell applications. *Int. J. Hydrog. Energy* **2010**, *35*, 9349–9384. [CrossRef]
6. Wang, Y.; Chen, K.S.; Mishler, J.; Cho, S.C.; Adroher, X.C. A review of polymer electrolyte membrane fuel cells: Technology, applications, and needs on fundamental research. *Appl. Energy* **2011**, *88*, 981–1007. [CrossRef]
7. Wan, Z.M.; Chang, H.W.; Shu, S.M.; Wang, Y.X.; Tang, H.L. A Review on Cold Start of Proton Exchange Membrane Fuel Cells. *Energies* **2014**, *7*, 3179–3203. [CrossRef]
8. Omrani, R.; Shabani, B. Gas diffusion layer modifications and treatments for improving the performance of proton exchange membrane fuel cells and electrolysers: A review. *Int. J. Hydrog. Energy* **2017**, *42*, 28515–28536. [CrossRef]
9. Leeuwner, M.J.; Patra, A.; Wilkinson, D.P.; Gyenge, E.L. Graphene and reduced graphene oxide based microporous layers for high-performance proton-exchange membrane fuel cells under varied humidity operation. *J. Power Sources* **2019**, *423*, 192–202. [CrossRef]
10. Park, S.; Lee, J.W.; Popov, B.N. A review of gas diffusion layer in PEM fuel cells: Materials and designs. *Int. J. Hydrog. Energy* **2012**, *37*, 5850–5865. [CrossRef]
11. Morgan, J.M.; Datta, R. Understanding the gas diffusion layer in proton exchange membrane fuel cells. I. How its structural characteristics affect diffusion and performance. *J. Power Sources* **2014**, *251*, 269–278. [CrossRef]
12. Mariani, M.; Latorrata, S.; Stampino, P.G.; Dotelli, G. Evaluation of Graphene Nanoplatelets as a Microporous Layer Material for PEMFC: Performance and Durability Analysis. *Fuel Cells* **2019**, *19*, 685–694. [CrossRef]
13. Chang, H.M.; Lin, C.W.; Chang, M.H.; Shiu, H.R.; Chang, W.C.; Tsau, F.H. Optimization of polytetrafluoroethylene content in cathode gas diffusion layer by the evaluation of compression effect on the performance of a proton exchange membrane fuel cell. *J. Power Sources* **2011**, *196*, 3773–3780. [CrossRef]

14. Gurau, V.; Bluemle, M.J.; De Castro, E.S.; Tsou, Y.M.; Mann, J.A.; Zawodzinski, T.A. Characterization of transport properties in gas diffusion layers for proton exchange membrane fuel cells—1. Wettability (internal contact angle to water and surface energy of GDL fibers). *J. Power Sources* **2006**, *160*, 1156–1162. [CrossRef]
15. Gurau, V.; Bluemle, M.J.; De Castro, E.S.; Tsou, Y.M.; Zawodzinski, T.A.; Mann, J.A. Characterization of transport properties in gas diffusion layers for proton exchange membrane fuel cells 2. Absolute permeability. *J. Power Sources* **2007**, *165*, 793–802. [CrossRef]
16. Wang, Y.; Wang, C.Y.; Chen, K.S. Elucidating differences between carbon paper and carbon cloth in polymer electrolyte fuel cells. *Electrochim. Acta* **2007**, *52*, 3965–3975. [CrossRef]
17. Ozden, A.; Shahgaldi, S.; Li, X.G.; Hamdullahpur, F. A graphene-based microporous layer for proton exchange membrane fuel cells: Characterization and performance comparison. *Renew Energy* **2018**, *126*, 485–494. [CrossRef]
18. Alink, R.; Gerteisen, D. Modeling the Liquid Water Transport in the Gas Diffusion Layer for Polymer Electrolyte Membrane Fuel Cells Using a Water Path Network. *Energies* **2013**, *6*, 4508–4530. [CrossRef]
19. Lee, J.; Liu, H.; George, M.G.; Banerjee, R.; Ge, N.; Chevalier, S.; Kotaka, T.; Tabuchi, Y.; Bazylak, A. Microporous layer to carbon fibre substrate interface impact on polymer electrolyte membrane fuel cell performance. *J. Power Sources* **2019**, *422*, 113–121. [CrossRef]
20. Latorrata, S.; Stampino, P.G.; Cristiani, C.; Dotelli, G. Performance Evaluation and Durability Enhancement of FEP-Based Gas Diffusion Media for PEM Fuel Cells. *Energies* **2017**, *10*, 2063. [CrossRef]
21. Kitahara, T.; Nakajima, H.; Mori, K. Hydrophilic and hydrophobic double microporous layer coated gas diffusion layer for enhancing performance of polymer electrolyte fuel cells under no-humidification at the cathode. *J. Power Sources* **2012**, *199*, 29–36. [CrossRef]
22. Leeuwner, M.J.; Wilkinson, D.P.; Gyenge, E.L. Novel Graphene Foam Microporous Layers for PEM Fuel Cells: Interfacial Characteristics and Comparative Performance. *Fuel Cells* **2015**, *15*, 790–801. [CrossRef]
23. Weber, A.Z.; Newman, J. Effects of microporous layers in polymer electrolyte fuel cells. *J. Electrochem. Soc.* **2005**, *152*, A677–A688. [CrossRef]
24. Park, S.; Popov, B.N. Effect of hydrophobicity and pore geometry in cathode GDL on PEM fuel cell performance. *Electrochim. Acta* **2009**, *54*, 3473–3479. [CrossRef]
25. Wong, A.K.C.; Ge, N.; Shrestha, P.; Liu, H.; Fahy, K.; Bazylak, A. Polytetrafluoroethylene content in standalone microporous layers: Tradeoff between membrane hydration and mass transport losses in polymer electrolyte membrane fuel cells. *Appl. Energy* **2019**, *240*, 549–560. [CrossRef]
26. Shrestha, P.; Ouellette, D.; Lee, J.; Ge, N.; Kai, A.; Wong, C.; Muirhead, D.; Liu, H.; Banerjee, R.; Bazylak, A. Graded Microporous Layers for Enhanced Capillary-Driven Liquid Water Removal in Polymer Electrolyte Membrane Fuel Cells. *Adv. Mater. Interfaces* **2019**, *6*, 1901157. [CrossRef]
27. Balzarotti, R.; Latorrata, S.; Stampino, P.G.; Cristiani, C.; Dotelli, G. Development and Characterization of Non-Conventional Micro-Porous Layers for PEM Fuel Cells. *Energies* **2015**, *8*, 7070–7083. [CrossRef]
28. Latorrata, S.; Balzarotti, R.; Stampino, P.G.; Cristiani, C.; Dotelli, G.; Guilizzoni, M. Design of properties and performances of innovative gas diffusion media for polymer electrolyte membrane fuel cells. *Prog. Org. Coat.* **2015**, *78*, 517–525. [CrossRef]
29. Yuan, X.Z.; Song, C.; Wang, H.; Zhang, J. *Electrochemical Impedance Spectroscopy in PEM Fuel Cells: Fundamentals and Applications*; Springer: London, UK, 2010; pp. 1–420.
30. Latorrata, S.; Pelosato, R.; Stampino, P.G.; Cristiani, C.; Dotelli, G. Use of Electrochemical Impedance Spectroscopy for the Evaluation of Performance of PEM Fuel Cells Based on Carbon Cloth Gas Diffusion Electrodes. *J. Spectrosc.* **2018**, *2018*, 3254375. [CrossRef]
31. Asghari, S.; Mokmeli, A.; Samavati, M. Study of PEM fuel cell performance by electrochemical impedance spectroscopy. *Int. J. Hydrog. Energy* **2010**, *35*, 9283–9290. [CrossRef]
32. Wagner, N. Characterization of membrane electrode assemblies in polymer electrolyte fuel cells using a.c. impedance spectroscopy. *J. Appl. Electrochem.* **2002**, *32*, 859–863. [CrossRef]
33. Dhirde, A.M.; Dale, N.V.; Salehfar, H.; Mann, M.D.; Han, T.H. Equivalent Electric Circuit Modeling and Performance Analysis of a PEM Fuel Cell Stack Using Impedance Spectroscopy. *IEEE Trans. Energy Convers.* **2010**, *25*, 778–786. [CrossRef]
34. Ramasamy, R.P.; Kumbur, E.C.; Mench, M.M.; Liu, W.; Moore, D.; Murthy, M. Investigation of macro- and micro-porous layer interaction in polymer electrolyte fuel cells. *Int. J. Hydrog. Energy* **2008**, *33*, 3351–3367. [CrossRef]

35. Latorrata, S.; Stampino, P.G.; Cristiani, C.; Dotelli, G. Development of an optimal gas diffusion medium for polymer electrolyte membrane fuel cells and assessment of its degradation mechanisms. *Int. J. Hydrog. Energy* **2015**, *40*, 14596–14608. [CrossRef]
36. Wang, X.L.; Zhang, H.M.; Zhang, J.L.; Xu, H.F.; Zhu, X.B.; Chen, J.; Yi, B.L. A bi-functional micro-porous layer with composite carbon black for PEM fuel cells. *J. Power Sources* **2006**, *162*, 474–479. [CrossRef]

© 2020 by the authors. Licensee MDPI, Basel, Switzerland. This article is an open access article distributed under the terms and conditions of the Creative Commons Attribution (CC BY) license (http://creativecommons.org/licenses/by/4.0/).

Article

Gas Diffusion Layers in Fuel Cells and Electrolysers: A Novel Semi-Empirical Model to Predict Electrical Conductivity of Sintered Metal Fibres

Reza Omrani and Bahman Shabani *

School of Engineering, RMIT University, Bundoora East Campus, Melbourne 3083, Australia; reza.omrani@rmit.edu.au
* Correspondence: bahman.shabani@rmit.edu.au; Tel.: +61-3-9925-4353

Received: 21 December 2018; Accepted: 27 February 2019; Published: 5 March 2019

Abstract: This paper introduces novel empirical as well as modified models to predict the electrical conductivity of sintered metal fibres and closed-cell foams. These models provide a significant improvement over the existing models and reduce the maximum relative error from as high as just over 30% down to about 10%. Also, it is shown that these models provide a noticeable improvement for closed-cell metal foams. However, the estimation of electrical conductivity of open-cell metal foams was improved marginally over previous models. Sintered porous metals are widely used in electrochemical devices such as water electrolysers, unitised regenerative fuel cells (URFCs) as gas diffusion layers (GDLs), and batteries. Having a more accurate prediction of electrical conductivity based on variation by porosity helps in better modelling of such devices and hence achieving improved designs. The models presented in this paper are fitted to the experimental results in order to highlight the difference between the conductivity of sintered metal fibres and metal foams. It is shown that the critical porosity (maximum achievable porosity) can play an important role in sintered metal fibres to predict the electrical conductivity whereas its effect is not significant in open-cell metal foams. Based on the models, the electrical conductivity reaches zero value at 95% porosity rather than 100% for sintered metal fibres.

Keywords: porous metal; porosity; sintered metal fibre; metal foam; electrical conductivity; electrochemical; fuel cell

1. Introduction

In recent decades, the development and use of porous metals in different engineering applications have increased substantially. Thanks to their highly practical characteristics, on top of bulk metals properties, these types of materials have functional and structural applications in a wide range of industries such as energy transportation and biomedical applications [1,2]. The main characteristics of porous metals are: low density, high permeability in open structures, high energy absorption and large specific surface area [1,3]. These properties that are linked with good electrical and thermal conductivity, corrosion resistance (when corrosion resistive metals such as titanium and stainless steel are used), and even distribution of the fluids, make them very attractive for electrochemical energy conversion devices [1,4]. For example, they are widely used as electrodes in batteries [1], gas diffusion layer (GDL) [4–6] or flow-field [7,8] in electrolysers, fuel cells and unitised regenerative fuel cells (URFCs). They can be either in the form of metal foams, sintered metallic powders/fibres or mesh.

Metal foams, commonly Ni, Pb and Cu, are mostly used as electrodes in batteries [9]; Ni foam has also been used in fuel cells as flow field [10,11]. Sintered metal powders/fibres (Figure 1) are commonly used as GDLs in polymer electrolyte membrane (PEM) and solid oxide (SO) electrolysers or URFCs [9,10] instead of carbon-based GDLs that usually corrode rapidly in such hydrated environments. Sintered metal fibres or powders are of particularly high interest due to their strong mechanical properties that make them suitable for providing mechanical support for other components in applications such as high-pressure water electrolysis [6,9,12]. Sintered metal powders provide porosities lower than 50% [13,14], which is suitable for electrolyser applications [15]; on the other hand, sintered metal fibres can provide porosities higher than 50% that is suitable for applications such as PEMFCs or URFCs [16,17]. In these kinds of applications, the electrical conductivity plays a major role in the performance as higher conductivity translates into lower ohmic losses, and hence higher efficiency.

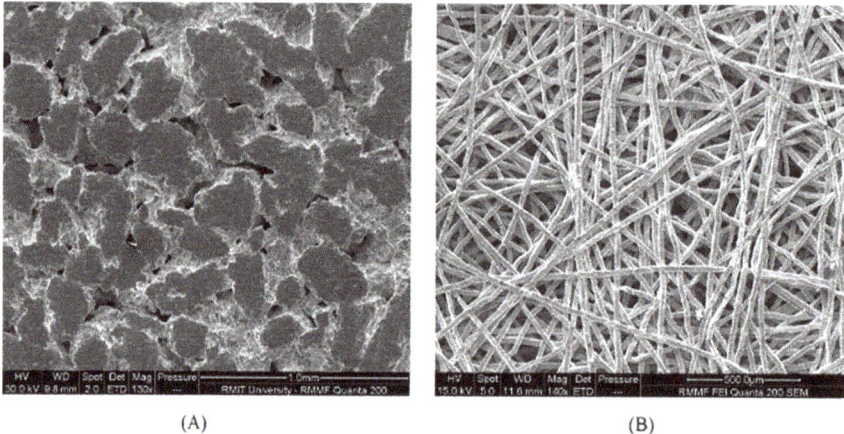

Figure 1. SEM images of sintered titanium powder (**A**) and sintered titanium fibre (**B**) commonly used as a gas diffusion layer (GDL) in unitised regenerative fuel cells (URFCs) and electrolysers.

In electrochemical applications, high electrical conductivity is preferable that is required for transferring the electrons effectively and hence, achieving higher efficiency. The electrical resistance of the porous metal (i.e., electrode) contributes to the ohmic losses. The electrical resistance of a porous metal increases by increasing its porosity; however, in many applications such as fuel cells, higher porosities are favourable for better delivery of reactants to the active area [17–23]. Therefore, the optimal porosity is a trade-off between the electrical conductivity and the species transport properties [18,24]. There are not many experimental data reported on the impact of GDL's porosity on the performance of fuel cells or other similar devices. Many conventional fuel cells use carbon-based GDL and such materials do not give enough control over parameters such as pore size and porosity. In addition, as these GDLs are compressible, the in-situ porosities are different from the ex-situ porosities, and they are significantly affected by the assembly pressure [25–27]. However, in URFCs and electrolysers, this issue does not exist as porous metallic GDLs are used with higher mechanical strength. Therefore, the in-situ properties of these GDLs remain almost similar to ex-situ properties, and they also give more opportunity to control different properties of the GDL and experimentally study their individual effect on the electrochemical device performance in a more controlled fashion.

Despite this fact, the variation of electrical conductivity with porosity is neglected in many simulations [20–22,28,29] as there are not reliable models to predict this property for fibrous materials and hence usually a constant value is considered. Therefore, it is of great importance to have models capable of predicting the electrical conductivity of porous metals, especially sintered porous metals, in order to perform more accurate simulations to predict the performance of electrochemical devices. Such a model can help determine the effect of parameters such as porosity of the porous electrode (e.g., in fuel cells, electrolysers, URFCs and batteries).

The electrical conductivity of porous metals is considerably lower than the bulk material in high porosity ranges. For example, based on measurements of electrical conductivity of aluminium foam by Feng et al. [30], the electrical conductivity at 70.5% and 86.7% porosity is approximately 1.708×10^6 m^{-1} and 0.5×10^6 S·m^{-1}, respectively, compared to 3.69×10^7 S·m^{-1} for aluminium. As a result, the electrical resistance of GDL increases rapidly at high porosities, which in turn translates into ohmic losses [31,32].

Several researchers have aimed to provide empirical relationships to link the electrical conductivity of the porous metal materials with the properties of the bulk metal and its porosity [30,33,34] or pore structure [35,36]. Feng et al. [30] have suggested a simplified model based on the model proposed by Huang [37]. Liu et al. [34,38,39] also conducted a series of experiments to determine the electrical resistivity of nickel foams developed by electroplating on the polyether sponge sheet using a double circuit bridge. They proposed a physical model for high porosity metal foams.

However, these formulas have been developed based on metal foams and their applicability for sintered metal fibres remains questionable. In these formulas (presented in Section 2.2), the porous metal electrical conductivity becomes zero when the porosity is 100% that cannot happen physically for sintered fibres or powders. High porosities close to 100% can be achieved in open-cell metal foams. However, in porous metals produced by sintering powders or fibres, the initial or maximum porosity (i.e., tap porosity before sintering) is lower than 100% (it can be increased by methods such as using pore-formers). Zhou et al. [40] have examined the applicability of three models, Equations (2)–(4), for sintered metal fibres; they have reported that these models, which are proposed for metal foams, cannot predict the results accurately as errors of more than 30% were observed. The electrical conductivity of sintered porous metals depends on several factors and the complex interaction between them such as impurities, manufacturing method, temperature, pressure and duration. The work of carried out by Sheng et al. [41] and Huang et al. [42] on thermal conductivity (analogous to electrical conductivity) of sintered metal fibres, highlights the complexity and difficulty in prediction of electrical conductivity of such products. It highlights the fact that there is a need for fitting parameters, detailed microstructural analysis [42], or reliance of measurement of other properties of the sintered metals (e.g., measurement of electrical conductivity to assist with the prediction of thermal conductivity [41]) to estimate the desired characteristic.

Accordingly, in this paper, it will be investigated whether the models can be improved to predict the electrical conductivity of sintered metal fibres with less error. The models analysed by Zhou et al. [40], are fitted to different experimental results from literature; they are modified considering the critical porosity to achieve higher accuracy. Also, two new models are presented for sintered metal fibres to improve the estimation of electrical conductivity. These models are also shown to improve the estimation of the electrical conductivity of closed-cell metal foams.

2. Method

2.1. Factors Affecting Electrical Conductivity of Porous Metals

Most of the models used for predicting the electrical conductivity of porous media are a function of porosity. Pore size has been shown to have a negligible effect on electrical conductivity [30,36,40,43]. Feng et al. [30] have analysed aluminium foams with pore diameters of 1.7, 2.5 and 3.6 mm and have reported a negligible difference between them. Zhou et al. [40] have analysed sintered copper fibres and have suggested that pore size has a minor effect on the electrical conductivity. Similarly, Goodall et al. [43] have reported no effect of pore size on the electrical conductivity of open cell aluminium foams; they have measured the electrical conductivity of aluminium foams with pore sizes of 75 μm and 400 μm with porosities between 64% and 93%. Hakamada et al. [36] have investigated the effect of pore size on electrical conductivity of porous aluminium (Figure 2): for the range that they have considered (i.e., from 212–300 to 850–1000 μm) the effect of pore size on electrical conductivity was found to be negligible, especially with pore sizes smaller than 600 μm.

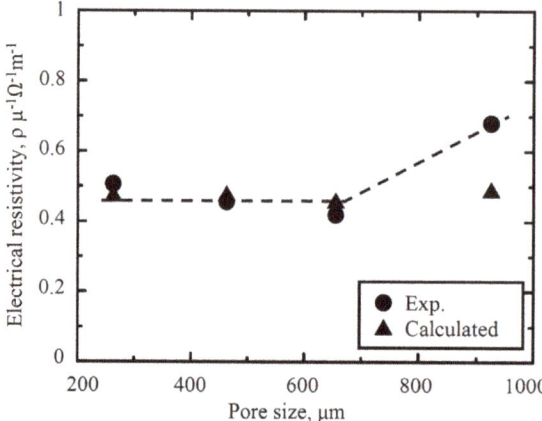

Figure 2. Effect of pore size on the electrical conductivity of porous aluminium with 80% porosity [36].

Some researchers have tried to predict the electrical conductivity according to the pore size and structure for porous metals [36]. Hakamada et al. [36] investigated porous aluminium produced by powder metallurgy and using spacers. They considered samples with a porosity of 77–90% and a pore diameter of 212–300 to 850–1000 μm. They proposed a model based on the porosity, pore size, and the size of openings in the cells. The model is in good agreement with the experimental results except for the pore size of 925 μm for which the model underestimates the experimental value by almost 30% and hence the range of its applicability should be investigated further. This model, however, can be used mainly for porous metals with controlled pore size and cannot be applied to samples with a range of random pore diameter.

As indicated by the literature the effect of pore size on electrical conductivity is negligible and this assumption offers more accuracy for pore sizes of lower than 100 μm (e.g., used in GDLs for fuel cell applications). Hence the focus of the models to be discussed in this paper is mainly on the effect of porosity (i.e., on electrical conductivity). In addition, models which are based on porosity can be easily applied to different samples without worrying about the pore size.

2.2. Porous Metal Electrical Conductivity Models

Several semi-empirical models have been proposed to estimate the electrical conductivity based on the conductivity of the bulk material and the porosity of porous metals [30,33,34]. Langlois and Coeuret [33] have proposed the following formula for highly porous metal foams:

$$\sigma = \frac{1-\varepsilon}{4}\sigma_0 \tag{1}$$

In this equation, σ is the electrical conductivity of the porous metal, ε is the porosity, and σ_0 is the conductivity of the bulk metal, which in this case is the metal. This model has been proposed based on the measurements of electrical conductivity of nickel foams with porosities in a very narrow range: 97.0% to 97.8%. This model is only applicable to the nickel foams with porosities in the aforementioned range which are made with specific conditions. Different manufacturing conditions and also defects in the samples can affect the electrical conductivity of porous metals [30,40]. In order to include these effects, Liu et al. [34,39] modified the Langlois and Coeuret formula by introducing a coefficient, K:

$$\sigma = K\frac{1-\varepsilon}{4}\cdot\sigma_0 \tag{2}$$

K represents the effect of different manufacturing methods and parameters, defects and dimensional variation of cells or particles and other possible factors affecting the electrical conductivity of a porous metal which results in deviation from theory. Its value is found through curve fitting to the measured data. They have shown that Equation (1) used for nickel foams with a porosity between 88% and 99% results in absolute relative errors (ARE) from 40% to 17%, respectively. It can be observed that as the porosity decreases the error of Equation (1) increases, whereas the maximum ARE from Equation (2) is 16.1% at 98.84% porosity.

Liu et al. [34,39] have also proposed a physical model based on the octahedron theoretical model for highly porous metal foams with porosities higher than 90%:

$$\sigma = K \frac{1-\varepsilon}{3[1 - 0.121 \cdot (1-\varepsilon)^{1/2}]} \cdot \sigma_0 \quad (3)$$

For the same range, this model has better accuracy and the maximum ARE is 14.9%. Both of these models have higher errors at higher porosities. This will be discussed further in Section 3.

Feng et al. [30] have proposed the following formula for calculating the electrical conductivity of closed cell metal foams:

$$\sigma = \frac{2K(1-\varepsilon)}{2K+\varepsilon} \cdot \sigma_0 \quad (4)$$

For spherical cells K has been suggested to be around 0.3.

Zhou et al. [40] have used the three models suggested by Liu et al. [34] and Feng et al. [30] for sintered copper fibres with porosities between 70% to 90%. The relative absolute error of the models for some points is as high as 30% (Table 1).

Table 1. Experimental data from Zhou et al. [40] and calculated values of electrical conductivity of sintered copper fibres using different models.

Porosity	Exp.	Electrical Conductivity (10⁶ S·m⁻¹)								Absolute Relative Error							
		Equation (2)		Equation (3)		Equation (4)				Equation (2)		Equation (3)		Equation (4)			
		Original	Equation (5)	Original	Equation (6)	Original	Equation (7)			Original	Equation (5)	Original	Equation (6)	Original	Equation (7)		
70%	0.291	0.216	0.286	0.247	0.291	0.219	0.287			26%	2%	15%	0%	25%	1%		
75%	0.228	0.180	0.228	0.192	0.220	0.181	0.228			21%	0%	16%	3%	21%	0%		
80%	0.144	0.144	0.170	0.144	0.158	0.144	0.170			0%	18%	0%	10%	0%	18%		
85%	0.102	0.108	0.113	0.102	0.103	0.107	0.112			6%	10%	0%	1%	5%	10%		
90%	0.055	0.072	0.055	0.064	0.054	0.071	0.055			31%	0%	17%	2%	29%	0%		

3. Improvement of the Existing Models

The discussed models in Section 2 predict that conductivity becomes zero when the porosity is 100%. In order to have electrical conductivity in a porous media, a connected solid phase is required. As can be seen in Figure 1, a porous metal is formed by randomly distributed pores of different sizes, which are formed by the surrounding complicated metallic structure (i.e., infinite cluster). The integrity of the porous metal relies on the existence of such a cluster. In the absence of a continuous cluster, the effective electrical conductivity and other properties become zero. Consequently, conductivity can become zero at porosities lower than 100%, which is the critical porosity for electrical conductivity or structural integrity. If the models are modified by introducing ε_c as the critical porosity, the models proposed by [34] and [30] will be as follows:

$$\sigma = K \frac{1 - \varepsilon/\varepsilon_c}{4} \cdot \sigma_0 \tag{5}$$

$$\sigma = K \frac{1 - \varepsilon/\varepsilon_c}{3[1 - 0.121 \cdot (1 - \varepsilon)^{1/2}]} \cdot \sigma_0 \tag{6}$$

$$\sigma = \frac{2K(1 - \varepsilon/\varepsilon_c)}{2K + \varepsilon} \cdot \sigma_0 \tag{7}$$

In these equations, the electrical conductivity will be zero when porosity is equal to the critical porosity, ε_c. These models are fitted to the experimental data from Feng et al. [30], Zhou et al. [40] and Liu et al. [34]. Feng et al. [30] have measured the electrical conductivity of closed-cell Al-alloy foams; Zhou et al. [40] have reported the electrical conductivity of sintered copper fibres; and Liu et al. [34] have investigated the electrical conductivity of open-cell nickel foams. It is expected that closed-cell foams have a lower critical porosity compared to open-cell foams. The reason is that more solid phase metal is required to form the structure of close-cell foams compared to that needed in open-cell porous metals.

Table 1 presents the experimental measurement of Zhou et al. [34] with the original equations and the modified models. The data are presented in Figure 3.

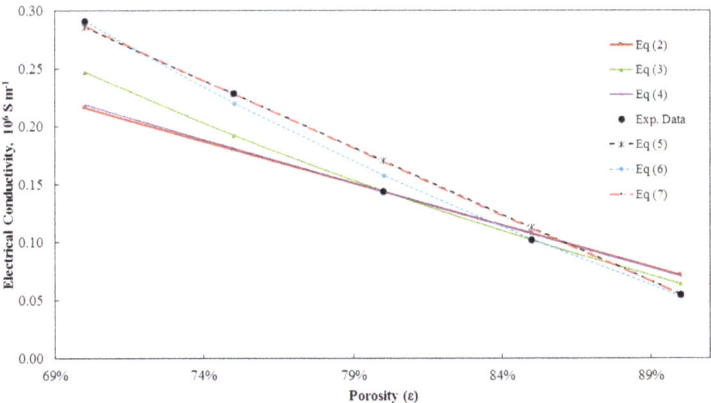

Figure 3. Different models fitted to experimental data from Zhou et al. [40].

As it can be seen in Table 1, the modified models result in considerably less error. The maximum ARE for Equations (5)–(7) is 18%, 10% and 18%, respectively, compared to 31%, 17% and 29% for Equations (2)–(4). The values of critical porosity for different models are calculated by fitting the models to the experimental data through minimising the ARE. The critical porosity for Equations (5)–(7) is 0.948, 0.962 and 0.949, respectively.

The critical porosity for the three fitted models is around 95% that is consistent with what is expected in reality. Equation (2) and its modified version, Equation (5), are presented against the experimental data from Feng et al. [30] in Figure 4. From this figure, it is evident that the experimental data do not approach zero conductivity at 100% porosity. The presumption of having zero conductivity and 100% porosity results in substantial errors and does not reflect an accurate relation between porosity and electrical conductivity.

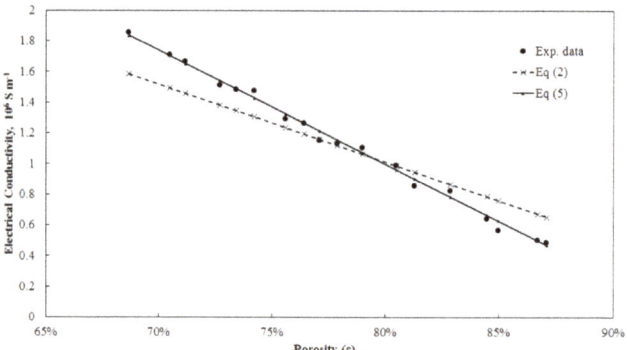

Figure 4. Prediction of experimental data from Feng et al. [30] by Equation (2) and its modified version, Equation (5).

Also, the models have been fitted to Feng et al. [30] measurements (Table 2) and a significant improvement has been observed. The maximum ARE has reduced from 35%, 26% and 34% for Equations (2)–(4), respectively, to 11%, 9% and 11% for Equations (5)–(7), respectively. Similarly, the average ARE has reduced from 13%, 7% and 12% to 3% for the modified models. From these results, it can be concluded that Equations (2)–(4) are not able to estimate the electrical conductivity at high porosities and considerable deviation can be observed. By introducing the critical porosity, the models perform significantly better. Therefore, it is concluded that the assumption of having zero electrical conductivity at 100% porosity or zero relative density cannot be valid. After fitting the models to these data, the critical porosity is 94% for Equations (5) and (7) and 95% for Equation (6).

The models were fitted to the experimental data from Liu et al. [34] that are presented in Figure 5. The models and their modified versions perform almost the same. The maximum ARE is between 12% and 17% for all models and the average ARE is between 5.5% and 8.2%. As it was expected, the critical porosity calculated from the modified models for these data was very close to 1 as the samples had an open-cell structure. The calculated critical porosity after fitting the models to the data was 99.7% for Equations (5) and (7) and 99.8% for Equation (6). As the critical porosity is close to unity, the improvement of the models is not significant.

Table 2. Experimental data from Feng et al. [30] and calculated values of electrical conductivity of Al-alloy foams using different models.

Porosity	Exp.	Electrical Conductivity (10^6 S m^{-1})						Absolute Relative Error					
		Equation (2)		Equation (3)		Equation (4)		Equation (2)		Equation (3)		Equation (4)	
		Original	Equation (5)	Original	Equation (6)	Original	Equation (7)	Original	Equation (5)	Original	Equation (6)	Original	Equation (7)
0.687	1.855	1.586	1.837	1.809	1.896	1.610	1.845	15%	1%	2%	2%	13%	1%
0.705	1.708	1.495	1.703	1.666	1.727	1.514	1.708	12%	0%	2%	1%	11%	0%
0.712	1.664	1.459	1.651	1.612	1.664	1.477	1.655	12%	1%	3%	0%	11%	1%
0.727	1.511	1.383	1.540	1.499	1.532	1.398	1.541	8%	2%	1%	1%	8%	2%
0.734	1.482	1.348	1.488	1.448	1.472	1.361	1.488	9%	0%	2%	1%	8%	0%
0.742	1.474	1.307	1.429	1.391	1.404	1.318	1.428	11%	3%	6%	5%	11%	3%
0.756	1.293	1.236	1.325	1.293	1.290	1.245	1.323	4%	2%	0%	0%	4%	2%
0.764	1.265	1.196	1.265	1.239	1.226	1.203	1.263	5%	0%	2%	3%	5%	0%
0.771	1.153	1.160	1.213	1.192	1.171	1.166	1.210	1%	5%	3%	2%	1%	5%
0.779	1.131	1.120	1.154	1.140	1.109	1.124	1.151	1%	2%	1%	2%	1%	2%
0.79	1.106	1.064	1.072	1.069	1.026	1.066	1.068	4%	3%	3%	7%	4%	3%
0.805	0.988	0.988	0.961	0.976	0.917	0.988	0.957	0%	3%	1%	7%	0%	3%
0.813	0.853	0.947	0.901	0.927	0.860	0.946	0.897	11%	6%	9%	1%	11%	5%
0.829	0.822	0.866	0.782	0.833	0.749	0.863	0.779	5%	5%	1%	9%	5%	5%
0.845	0.639	0.785	0.663	0.742	0.642	0.781	0.661	23%	4%	16%	0%	22%	3%
0.85	0.565	0.760	0.626	0.714	0.609	0.755	0.624	35%	11%	26%	8%	34%	11%
0.867	0.500	0.674	0.500	0.622	0.500	0.667	0.500	35%	0%	24%	0%	33%	0%
0.871	0.486	0.654	0.470	0.600	0.475	0.647	0.471	34%	3%	24%	2%	33%	3%

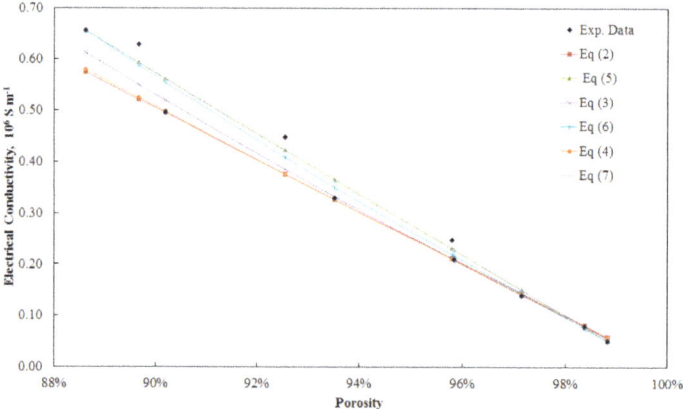

Figure 5. Fitting of different models to experimental data from Liu et al. [34].

4. New empirical Models for Prediction of Electrical Conductivity of Porous Metals

4.1. An Overview

As discussed, the existing models fail to give reliable predictions of electrical conductivity for sintered metal fibres and as can be seen in Section 3, errors can be as high as 30%. Modified models improved the accuracy of the existing models. It was found out that the main reason for high inaccuracy is neglecting the critical porosity. In this section new models will be proposed to improve the accuracy of these existing models. The models will include the critical porosity as well as the tortuosity effect on electrical conductivity.

4.2. Models to Predict Sintered Metal Fibres Electrical Conductivity

In this section according to the rule of mixture, and considering the effect of tortuosity and critical porosity, two new models for predicting the electrical conductivity of porous materials are proposed. Based on the rule of mixture, the electrical conductivity of unidirectional metal fibres can be expressed as below [44]:

$$\sigma = \varepsilon \sigma_{air} + (1-\varepsilon)\sigma_s \quad (8)$$

Considering that the electrical conductivity of air is negligible compared to the conductivity of the solid metal phase, therefore the electrical conductivity can be expressed as:

$$\sigma = \sigma_s(1-\varepsilon) \quad (9)$$

Although this formula gives the correct value for long unidirectional fibres, this is not valid in sintered metal fibres as the length of the fibres can vary as well as their directions. Also, the connection of the fibres depends on the sintering conditions such as pressure, temperature, and duration [40]. For instance, Zhou et al. [40] have shown that for sintered copper fibres samples with a porosity of 90% porosity, the electrical conductivity at 1000 °C was almost 5.4 times as high as of the electrical conductivity at 700 °C. Therefore, in order to introduce the effect of manufacturing processes, inhomogeneity in the structure, and possible defects we can introduce a coefficient that accounts for these effects:

$$\sigma = K\sigma_s(1-\varepsilon) \quad (10)$$

It is expected that the value of K to be lower than one (i.e., K = 1 represents ideal conditions). Here the critical porosity, ε_c, can be introduced to the previous equation:

$$\sigma = K\sigma_s(1 - \varepsilon/\varepsilon_c) \qquad (11)$$

One factor that contributes to the increase of the conductivity in porous metals is the increase in the path that electrons need to travel, i.e., the electrical tortuosity, due to the presence of nonconductive voids. This is illustrated in Figure 6. It can be seen that the electrons have to travel a longer path in a porous metal in comparison to a bulk metal where they can travel in a straight line.

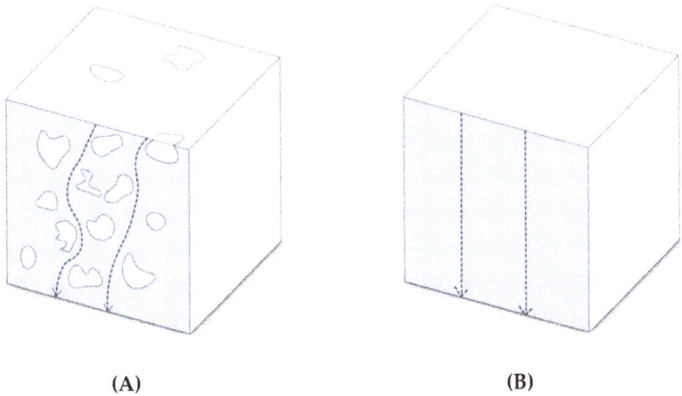

Figure 6. Electron path in (**A**) a porous and (**B**) a bulk metal.

This can be expressed by the tortuosity concept. If we assume in the bulk material, the length travelled by electrons is L_0 and in the porous material it is L, and consider τ as tortuosity, then we can write:

$$L = \tau L_0 \qquad (12)$$

This can also account for the fact that fibres are not unidirectional. The Bruggeman relation is a widely accepted estimation for tortuosity [45,46]:

$$\tau = \varepsilon^{-\alpha} \qquad (13)$$

However, this expression is hydraulic tortuosity in the void space in the porous media. α is a coefficient, which depends on the structure of the porous media. For the electric tortuosity, we can rewrite the Bruggeman relation as below as the solid phase is important:

$$\tau = (1 - \varepsilon)^{-\alpha} \qquad (14)$$

This relation satisfies the boundary conditions for tortuosity at $\varepsilon = 0$ and $\varepsilon = 1$. When porosity is equal to zero, the tortuosity must be 1 and when porosity is 1, tortuosity must be ∞.

Another accepted model for tortuosity is [47]:

$$\tau = 1 - P\ln(\varepsilon) \qquad (15)$$

Here, P is a fitting parameter that depends on the porous media structure and can be estimated experimentally or numerically, such as using lattice Boltzmann simulation. Again, to satisfy the following conditions: $\tau \to \infty$ when $\varepsilon = 1$ and $\tau \to 1$ when $\varepsilon = 0$, we can change the formula to:

$$\tau = 1 - P\ln(1 - \varepsilon) \qquad (16)$$

Now, by introducing the effect of tortuosity in Equation (10), we have:

$$\sigma = K\sigma_s(1 - \varepsilon/\varepsilon_c)\, \tau^{-1} \qquad (17)$$

By replacing tortuosity in Equation (17) by Equations (13) and (14) we have:

$$\sigma = K\sigma_s(1 - \varepsilon/\varepsilon_c)\, (1-\varepsilon)^\alpha \qquad (18)$$

$$\sigma = \frac{K\sigma_s(1 - \varepsilon/\varepsilon_c)}{1 - Pln(\varepsilon)} \qquad (19)$$

The results for the two new models are presented in Table 3 for experimental data from Zhou et al. [40] and in Table 4, for data from Feng et al. [30]. These models result in lower ARE compared to previous models. The two new models have almost similar results. The ARE for different structures and using different models can be seen in Table 5. It is clear that for sintered metal fibres the accuracy of the models can be significantly improved by the critical porosity introduction; the improvement is less for closed-cell metal foam; for the open-cell foam the improvement is insignificant.

Table 3. Experimental data from Zhou et al. [40] and calculated values of electrical conductivity of sintered copper fibres using the newly proposed models.

Porosity	Electrical Conductivity (10^6 S·m^{-1})			Absolute Relative Error	
	Exp.	Equation (18)	Equation (19)	Equation (18)	Equation (19)
70%	0.291	0.291	0.291	0.1%	0.0%
75%	0.228	0.221	0.220	3.0%	3.4%
80%	0.144	0.158	0.157	9.6%	9.1%
85%	0.102	0.102	0.102	0.0%	0.0%
90%	0.055	0.055	0.055	0.1%	0.0%

Table 4. Experimental data from Feng et al. [30] and calculated values of electrical conductivity of aluminum foam using the new proposed models.

Porosity	Electrical Conductivity (10^6 S m^{-1})			Absolute Relative Error	
	Exp.	Equation (18)	Equation (19)	Equation (18)	Equation (19)
68.70%	1.855	1.856	1.855	0.0%	0.0%
70.50%	1.708	1.710	1.709	0.1%	0.1%
71.20%	1.664	1.653	1.653	0.6%	0.7%
72.70%	1.511	1.534	1.533	1.5%	1.5%
73.40%	1.482	1.478	1.478	0.3%	0.3%
74.20%	1.474	1.415	1.415	4.0%	4.0%
75.60%	1.293	1.306	1.307	1.0%	1.0%
76.40%	1.265	1.245	1.245	1.6%	1.6%
77.10%	1.153	1.191	1.192	3.3%	3.3%
77.90%	1.131	1.130	1.131	0.0%	0.0%
79.00%	1.106	1.048	1.048	5.3%	5.2%
80.50%	0.988	0.937	0.937	5.2%	5.1%
81.30%	0.853	0.878	0.879	2.9%	3.0%
82.90%	0.822	0.763	0.764	7.2%	7.1%
84.50%	0.639	0.650	0.651	1.8%	1.9%
85.00%	0.565	0.616	0.616	9.0%	9.1%
86.70%	0.500	0.500	0.500	0.0%	0.0%
87.10%	0.486	0.473	0.473	2.6%	2.6%

Table 5. Average absolute relative error from different models for different structures.

Structure	Previous Models			Modified Models			New Models	
	Equation (2)	Equation (3)	Equation (4)	Equation (5)	Equation (6)	Equation (7)	Equation (18)	Equation (19)
Sintered metal fibre (Zhou et al.)	16.7%	9.6%	15.8%	6.1%	3.2%	5.8%	2.6%	2.5%
Closed-cell metal foam (Feng et al.)	12.6%	7.1%	11.9%	2.8%	2.9%	2.8%	2.6%	2.6%
Open-cell metal foam (Liu et al.)	8.19%	6.50%	7.59%	6.10%	5.54%	5.98%	5.44%	5.41%

Therefore, by looking at the results from the modified models and the new proposed models, it can be concluded that for sintered metal fibres, the presumption of having zero conductivity at a porosity of 100% does not provide accurate results. By doing so, the errors for sintered copper fibres can reach up to 30% for the range of 70%–90%. Also, as it was expected, the critical porosity obtained by fitting the models to the experimental results for open-cell metal foam, was higher (close to unity) than that of obtained for sintered metal fibres and closed-cell metal foam. The suggested reason is that for having structural integrity for a similar structure, closed-cell metal foam requires more solid phase. The average ARE from using different models for different structures can be seen in Table 5. As it can be observed, for the sintered metal fibres, significant improvement can be gained by the modified models and even more significantly by the new proposed models. For closed-cell the improvement is moderate; and finally, for the open cell metal foam, the reduction in error is marginal and average ARE is reduced by around 1% to 3%. As a result, it is suggested that for open cell metal foam Equation (3) can be used, which gives less error compared to Equations (2) and (4) and the inclusion of critical porosity does not bring significant improvement. For closed cell metal foams, modified models (i.e., Equations (5)–(7)) and new suggested models (i.e., Equations (18) and (19)) provide similar accuracy and outperform models without considering critical porosity. Finally, for sintered metal fibre porous media, the newly proposed models outperform the previous models and hence are recommended to be considered for this type of porous metal.

The critical porosity at which the electrical conductivity reaches zero (i.e., the maximum porosity that is achievable to have a connected cluster of solid phased) depends on the type of the porous metal (the method by which the porous metal is manufactured). The highest amount of porosity in porous metals can be achieved in metal foams, specifically open-cell metal foams. Metal foams typically have larger pore size compared to that of sintered metal fibres or powders. Larger pores mean higher void space ratio to solid phase. Also, in metal foams that are manufactured from molten metal, the solid phase is uniform and there is no contact resistance in the structure. Additionally, open-cell metal foams can provide higher porosities in comparison to the closed-cell metal foams as the walls (solid phase) are non-existent in open-cells and only struts are present to form a cell. On the other hand, sintered porous metals, are formed by means of pressure at high temperatures. This puts a limit on the maximum porosity achievable. Additionally, as the sintering depends on the temperature, pressure, and the duration, the electrical conductivity depends on these factors [40,48]. These factors affect the contact between the metal fibres or powders and hence the overall electrical conductivity. As a result, the critical porosity, has a more significant effect on the accuracy of the models for porous metals formed by sintering, followed by closed-cell metal foams, and finally less impact on the open-cell metal foams. This is also evident in the results by fitting the models and the critical porosity for the open-cell foam was obtained to be close to one, whereas for sintered metal fibres, this value was ~95%.

In order, to find the validity of the models for other cases of closed-cell metal foams the models are fitted to additional data from the literature. For metal foams, the data from Feng et al. [30], Sevostianov et al. [49], Kim et al. [50] and Kovacik et al. [51] are used. After fitting the data to these models, the coefficients are obtained as presented in Table 5. The critical porosity and the tortuosity coefficient are similar for different datasets. The value of K (manufacturing effects) is close for the three of the datasets ([30,50,51], However, K for the data from Sevostianov et al. [49] is slightly lower. The value of K is expected to vary between different datasets provided by different research groups as it depends on several factors and the complex interaction between them such as impurities and manufacturing methods implemented. If a K value based on these data is applied to a new dataset, a higher error could potentially arise due to the effect of aforementioned parameters. Therefore, it is suggested for new porous metals, few measurements of the electrical conductivity of samples with different porosities to be experimentally measured (e.g., two different porosities). Subsequently, the value of K can be determined based on those values to minimise the error and predict the electrical conductivity of the porous metal at different porosities by the model. This allows the prediction of the conductivity for a wide range of porosity that otherwise needs to be measured individually and samples with different porosities are required to be manufactured that are not practical in many circumstances. This will help in modelling of devices implementing such materials, where electrical conductivity is required such as fuel cells and electrolysers.

Also, if the value of K for closed metal foams (Table 6) is compared the value for sintered metal fibres (experimental data from Zhou et al. [40]), ~0.03, it is evident that the value of K is significantly higher for closed-cell metal foams as expected. This can be explained by the fact that in metal foams, the solid phase can be considered one connected piece, whereas in sintered porous metals, there are numerous connections points between the particles adding to the overall resistance (i.e., lower conductivity) resulting in a lower value of K. In other words, the value of K represents the deviation from ideal conditions.

Table 6. Coefficient of the new models: Equations (18) and (19), for closed-cell metal foams.

Data Sets	Equation (18)			Equation (19)		
	K	ε_c	α	K	ε_c	P
Feng et al.	0.80			0.81		
Kovacik et al.	0.76	0.99	0.32	0.77	0.97	0.37
Sevostianov et al.	0.67			0.67		
Kim et al.	0.78			0.79		

5. Conclusions

The effect of porosity on the electrical conductivity of different types of porous metals was investigated through different existing semi-empirical models and the applicability of these models was evaluated for open and closed cell metal foams as well as metal fibres. These models are proposed based on metal foams; that is why they can represent the electrical conductivity of open-cell metal foams with low errors for porosities over 80%. However, the error increases for closed-cell metal foams. For sintered metal fibres the error is even more considerable as it can be as high as 30%. These existing models that are predicting the conductivity to be zero at 100% porosity do not match the experimental measurements. In practice, porous metals with porosities close to 100% cannot be produced, especially if sintering technology is used, and the maximum achievable porosity is the tap porosity. Therefore, the critical porosity is introduced in the models. This means that the electrical conductivity should reach the value of zero at a porosity lower than 100% (i.e., critical porosity). This modification improves the models predictions significantly (i.e., reduces the error of the models). Moreover, two new empirical models (Equations (18) and (19)) based on mixture model, tortuosity, and critical porosity were proposed for sintered metal fibres. Using these models, the critical porosity (the porosity at which the electrical conductivity becomes zero) was estimated to be ~95% for sintered metal fibres rather than 100%. These models offered more accuracy and lower errors compared to previous models, especially for sintered metal fibres. The model proposed by Liu et al. [34,39], Equation (3), outperforms the other models, Equations (2) and (4), and its modified model, Equation (6) performed better than the modified version of the other models.

The two new models (Equations (18) and (19)) were also fitted to more experimental data from literature, in order to investigate if the values for the coefficients can be generalised based on the type of the porous metal. As currently, most of the data available in the literature are for closed-cell metal foams; hence, the present study was only carried out for this type of porous metals. It was found that the values of critical porosity and tortuosity coefficient can be applied to different datasets, however, the coefficient considered for manufacturing effect, K, is different for one of the datasets while the difference for other datasets are not significant.

In summary, it is suggested that for high porosity sintered metal fibres, which are usually used in electrochemical devices, the critical porosity (i.e., the maximum possible physical porosity) should be considered as it noticeably affects the accuracy of the models. By using the modified models and the new models, the average ARE was reduced from as high as 16.7% to as low as 2.5%. In addition, the maximum ARE in the unmodified models is from 17% to 31.8%, which was reduced to 9.1% to 18.3% by modifications and introduction of new models. Considering that these findings were based on the porous metals with porosities above 70%, further investigation is still required to check the applicability of these formulas for lower ranges of porosities, i.e., lower than 50%. It is worthwhile to mention that as contact resistance can significantly affect the ohmic losses in electrochemical devices the effect of porosity on contact resistance is recommended to be investigated separately.

Author Contributions: Conceptualisation, R.O.; methodology, R.O.; literature review, R.O.; validation, R.O. and B.S.; formal analysis, R.O. and B.S.; investigation, R.O. and B.S.; writing-original and draft preparation, R.O. writing-review and editing, R.O. and B.S.; project supervision, B.S.; and project administration R.O.

Funding: This research received no external funding.

Acknowledgments: The authors gratefully acknowledge the support through the provision of an Australian Government Research Training Program Scholarship by RMIT University.

Conflicts of Interest: The authors declare no conflict of interest.

References

1. Liu, P.S.; Chen, G.F. *Porous Materials: Processing and Applications*; Butterworth-Heinemann: Boston, MA, USA, 2014. [CrossRef]
2. Dukhan, N. *Metal Foams: Fundamentals and Applications*; Destech Publications: Lancaster, PA, USA, 2013. [CrossRef]
3. Lefebvre, L.P.; Banhart, J.; Dunand, D.C. Porous Metals and Metallic Foams: Current Status and Recent Developments. *Adv. Eng. Mater.* **2008**, *10*, 775–787. [CrossRef]
4. Sajid Hossain, M.; Shabani, B. Metal foams application to enhance cooling of open cathode polymer electrolyte membrane fuel cells. *J. Power Sources* **2015**, *295*, 275–291. [CrossRef]
5. Gabbasa, M.; Sopian, K.; Fudholi, A.; Asim, N. A review of unitized regenerative fuel cell stack: Material, design and research achievements. *Int. J. Hydrog. Energy* **2014**, *39*, 17765–17778. [CrossRef]
6. Carmo, M.; Fritz, D.L.; Mergel, J.; Stolten, D. A comprehensive review on PEM water electrolysis. *Int. J. Hydrog. Energy* **2013**, *38*, 4901–4934. [CrossRef]
7. Sajid Hossain, M.; Shabani, B. Air flow through confined metal foam passage: Experimental investigation and mathematical modelling. *Exp. Therm. Fluid Sci.* **2018**, *99*, 13–25. [CrossRef]
8. Sajid Hossain, M.; Shabani, B. Experimental study on confined metal foam flow passage as compact heat exchanger surface. *Int. Commun. Heat Mass Transf.* **2018**, *98*, 286–296. [CrossRef]
9. Liu, P.S.; Chen, G.F. Chapter Three—Application of Porous Metals. In *Porous Materials*; Chen, P.S.L.F., Ed.; Butterworth-Heinemann: Boston, MA, USA, 2014. [CrossRef]
10. Yuan, W.; Tang, Y.; Yang, X.; Wan, Z. Porous metal materials for polymer electrolyte membrane fuel cells—A review. *Appl. Energy* **2012**, *94*, 309–329. [CrossRef]
11. Arisetty, S.; Prasad, A.K.; Advani, S.G. Metal foams as flow field and gas diffusion layer in direct methanol fuel cells. *J. Power Sources* **2007**, *165*, 49–57. [CrossRef]
12. Tang, H.P.; Wang, J.; Qian, M. 28-Porous titanium structures and applications. In *Titanium Powder Metallurgy*; Qian, M., Froes, F.H., Eds.; Butterworth-Heinemann: Boston, MA, USA, 2015. [CrossRef]

13. Güden, M.; Çelik, E.; Hızal, A.; Altındiş, M.; Çetiner, S. Effects of compaction pressure and particle shape on the porosity and compression mechanical properties of sintered Ti6Al4V powder compacts for hard tissue implantation. *J. Biomed. Mater. Res. Part B Appl. Biomater.* **2008**, *85B*, 547–555. [CrossRef] [PubMed]
14. Grootenhuis, P.; Powell, R.W.; Tye, R.P. Thermal and Electrical Conductivity of Porous Metals made by Powder Metallurgy Methods. *Proc. Phys. Soc. Sect. B* **1952**, *65*, 502. [CrossRef]
15. Grigoriev, S.A.; Millet, P.; Volobuev, S.A.; Fateev, V.N. Optimization of porous current collectors for PEM water electrolysers. *Int. J. Hydrog. Energy* **2009**, *34*, 4968–4973. [CrossRef]
16. Hwang, C.M.; Ishida, M.; Ito, H.; Maeda, T.; Nakano, A.; Hasegawa, Y.; Yokoi, N.; Kato, A.; Yoshida, T. Influence of properties of gas diffusion layers on the performance of polymer electrolyte-based unitized reversible fuel cells. *Int. J. Hydrog. Energy* **2011**, *36*, 1740–1753. [CrossRef]
17. Omrani, R.; Shabani, B. Review of gas diffusion layer for proton exchange membrane-based technologies with a focus on unitised regenerative fuel cells. *Int. J. Hydrog. Energy* **2019**, *44*, 3834–3860. [CrossRef]
18. Lin, H.-H.; Cheng, C.-H.; Soong, C.-Y.; Chen, F.; Yan, W.-M. Optimization of key parameters in the proton exchange membrane fuel cell. *J. Power Sources* **2006**, *162*, 246–254. [CrossRef]
19. Cheema, T.A.; Zaidi, S.M.J.; Rahman, S.U. Three dimensional numerical investigations for the effects of gas diffusion layer on PEM fuel cell performance. *Renew. Energy* **2011**, *36*, 529–535. [CrossRef]
20. Maslan, N.H.; Gau, M.M.; Masdar, M.S.; Rosli, M.I. Simulation of porosity and PTFE content in gas diffusion layer on proton exchange membrane fuel cell performance. *J. Eng. Sci. Technol.* **2016**, *11*, 85–95.
21. Larbi, B.; Alimi, W.; Chouikh, R.; Guizani, A. Effect of porosity and pressure on the PEM fuel cell performance. *Int. J. Hydrog. Energy* **2013**, *38*, 8542–8549. [CrossRef]
22. Sahraoui, M.; Kharrat, C.; Halouani, K. Two-dimensional modeling of electrochemical and transport phenomena in the porous structures of a PEMFC. *Int. J. Hydrog. Energy* **2009**, *34*, 3091–3103. [CrossRef]
23. Mason, T.J.; Millichamp, J.; Shearing, P.R.; Brett, D.J.L. A study of the effect of compression on the performance of polymer electrolyte fuel cells using electrochemical impedance spectroscopy and dimensional change analysis. *Int. J. Hydrog. Energy* **2013**, *38*, 7414–7422. [CrossRef]
24. Omrani, R.; Shabani, B. Gas diffusion layer modifications and treatments for improving the performance of proton exchange membrane fuel cells and electrolysers: A review. *Int. J. Hydrog. Energy* **2017**, *42*, 28515–28536. [CrossRef]
25. Ye, D.; Gauthier, E.; Benziger, J.B.; Pan, M. Bulk and contact resistances of gas diffusion layers in proton exchange membrane fuel cells. *J. Power Sources* **2014**, *256*, 449–456. [CrossRef]
26. Xing, X.Q.; Lum, K.W.; Poh, H.J.; Wu, Y.L. Optimization of assembly clamping pressure on performance of proton-exchange membrane fuel cells. *J. Power Sources* **2010**, *195*, 62–68. [CrossRef]
27. Mason, T.J.; Millichamp, J.; Neville, T.P.; El-kharouf, A.; Pollet, B.G.; Brett, D.J.L. Effect of clamping pressure on ohmic resistance and compression of gas diffusion layers for polymer electrolyte fuel cells. *J. Power Sources* **2012**, *219*, 52–59. [CrossRef]
28. Obayopo, S.O.; Bello-Ochende, T.; Meyer, J.P. Three-dimensional optimisation of a fuel gas channel of a proton exchange membrane fuel cell for maximum current density. *Int. J. Energy Res.* **2013**, *37*, 228–241. [CrossRef]
29. Yan, W.M.; Soong, C.Y.; Chen, F.; Chu, H.S. Effects of flow distributor geometry and diffusion layer porosity on reactant gas transport and performance of proton exchange membrane fuel cells. *J. Power Sources* **2004**, *125*, 27–39. [CrossRef]
30. Feng, Y.; Zheng, H.; Zhu, Z.; Zu, F. The microstructure and electrical conductivity of aluminum alloy foams. *Mater. Chem. Phys.* **2003**, *78*, 196–201. [CrossRef]
31. Morgan, J.M.; Datta, R. Understanding the gas diffusion layer in proton exchange membrane fuel cells. I. How its structural characteristics affect diffusion and performance. *J. Power Sources* **2014**, *251*, 269–278. [CrossRef]
32. Netwall, C.J.; Gould, B.D.; Rodgers, J.A.; Nasello, N.J.; Swider-Lyons, K.E. Decreasing contact resistance in proton-exchange membrane fuel cells with metal bipolar plates. *J. Power Sources* **2013**, *227*, 137–144. [CrossRef]
33. Langlois, S.; Coeuret, F. Flow-through and flow-by porous electrodes of nickel foam. I. Material characterization. *J. Appl. Electrochem.* **1989**, *19*, 43–50. [CrossRef]
34. Liu, P.S.; Li, T.F.; Fu, C. Relationship between electrical resistivity and porosity for porous metals. *Mater. Sci. Eng. A* **1999**, *268*, 208–215. [CrossRef]

35. Dharmasena, K.P.; Wadley, H.N.G. Electrical Conductivity of Open-cell Metal Foams. *J. Mater. Res.* **2002**, *17*, 625–631. [CrossRef]
36. Hakamada, M.; Kuromura, T.; Chen, Y.; Kusuda, H.; Mabuchi, M. Influence of Porosity and Pore Size on Electrical Resistivity of Porous Aluminum Produced by Spacer Method. *Mater. Trans.* **2007**, *48*, 32–36. [CrossRef]
37. Huang, P.Y. *Principles of Powder Metallurgy*; Metallurgical Industry Press: Beijing, China, 1997.
38. Liu, P.; Fu, C.; Li, T. Calculation formula for apparent electrical resistivity of high porosity metal materials. *Sci. China Ser. E-Technol. Sci.* **1999**, *42*, 294–301. [CrossRef]
39. Liu, P.S.; Liang, K.M. Evaluating electrical resistivity for high porosity metals. *Mater. Sci. Technol.* **2000**, *16*, 341–343. [CrossRef]
40. Zhou, W.; Tang, Y.; Song, R.; Jiang, L.; Hui, K.S.; Hui, K.N. Characterization of electrical conductivity of porous metal fiber sintered sheet using four-point probe method. *Mater. Des.* **2012**, *37*, 161–165. [CrossRef]
41. Sheng, M.; Cahela, D.R.; Yang, H.; Gonzalez, C.F.; Yantz, W.R.; Harris, D.K.; Tatarchuk, B.J. Effective thermal conductivity and junction factor for sintered microfibrous materials. *Int. J. Heat Mass Transf.* **2013**, *56*, 10–19. [CrossRef]
42. Huang, X.; Zhou, Q.; Liu, J.; Zhao, Y.; Zhou, W.; Deng, D. 3D stochastic modeling, simulation and analysis of effective thermal conductivity in fibrous media. *Powder Technol.* **2017**, *320*, 397–404. [CrossRef]
43. Goodall, R.; Weber, L.; Mortensen, A. The electrical conductivity of microcellular metals. *J. Appl. Phys.* **2006**, *100*, 044912. [CrossRef]
44. Li, L.; Chung, D.D.L. Electrical and mechanical properties of electrically conductive polyethersulfone composites. *Composites* **1994**, *25*, 215–224. [CrossRef]
45. Ebner, M.; Wood, V. Tool for Tortuosity Estimation in Lithium Ion Battery Porous Electrodes. *J. Electrochem. Soc.* **2015**, *162*, A3064–A3070. [CrossRef]
46. Farmer, J.; Duong, B.; Seraphin, S.; Shimpalee, S.; Martínez-Rodríguez, M.J.; Van Zee, J.W. Assessing porosity of proton exchange membrane fuel cell gas diffusion layers by scanning electron microscope image analysis. *J. Power Sources* **2012**, *197*, 1–11. [CrossRef]
47. Saomoto, H.; Katagiri, J. Direct comparison of hydraulic tortuosity and electric tortuosity based on finite element analysis. *Theor. Appl. Mech. Lett.* **2015**. [CrossRef]
48. Kostornov, A.G.; Shevchuk, M.S.; Lezhenin, F.F.; Fedorchenko, I.M. An experimental investigation into the thermal and electrical conductivities of metal fiber materials. *Sov. Powder Metall. Metal Ceram.* **1977**, *16*, 194–197. [CrossRef]
49. Sevostianov, I.; Kováčik, J.; Simančík, F. Elastic and electric properties of closed-cell aluminum foams: Cross-property connection. *Mater. Sci. Eng. A* **2006**, *420*, 87–99. [CrossRef]
50. Kim, A.; Hasan, M.A.; Nahm, S.H.; Cho, S.S. Evaluation of compressive mechanical properties of Al-foams using electrical conductivity. *Comp. Struct.* **2005**, *71*, 191–198. [CrossRef]
51. Kováčik, J.; Simančík, F. Aluminium foam—Modulus of elasticity and electrical conductivity according to percolation theory. *Scr. Mater.* **1998**, *39*, 239–246. [CrossRef]

© 2019 by the authors. Licensee MDPI, Basel, Switzerland. This article is an open access article distributed under the terms and conditions of the Creative Commons Attribution (CC BY) license (http://creativecommons.org/licenses/by/4.0/).

Article

Innovative Membrane Electrode Assembly (MEA) Fabrication for Proton Exchange Membrane Water Electrolysis

Guo-Bin Jung *, Shih-Hung Chan, Chun-Ju Lai, Chia-Chen Yeh and Jyun-Wei Yu

Department of Mechanical Engineering of Yuan Ze University, Taoyuan City 32003, Taiwan; janshun@saturn.yzu.edu.tw (S.-H.C.); laichenru@yahoo.com.tw (C.-J.L.); nicdoit770212@gmail.com (C.-C.Y.); s1048703@g.yzu.edu.tw (J.-W.Y.)
* Correspondence: guobin@saturn.yzu.edu.tw; Tel.: +886-348638800-2469

Received: 1 October 2019; Accepted: 1 November 2019; Published: 5 November 2019

Abstract: In order to increase the hydrogen production rate as well as ozone production at the anode side, increased voltage application and more catalyst utilization are necessary. The membrane electrode assembly (MEA) produces hydrogen/ozone via proton exchange membrane water electrolysis (PEMWE)s which gives priority to a coating method (abbreviation: ML). However, coating takes more effort and is labor-consuming. This study will present an innovative preparation method, known as flat layer (FL), and compare it with ML. FL can significantly reduce efforts and largely improve MEA production. Additionally, MEA with the FL method is potentially durable compared to ML.

Keywords: proton exchange membrane water electrolysis (PEMWE), catalyst-coated membrane; hydrogen generation; membrane electrode assembly (MEA), ozone production; flat layer (FL)

---

1. Introduction

The greenhouse effect of this century is increasing and hence, a reduction of carbon dioxide emission through various methods is the global consensus. Developing clean and renewable energy has been the main target until now. However, the major issue of renewable energy is the unstable output of power that is affected by seasonal and environmental factors, which results in electric grid management difficulties. The advantages of proton exchange membrane water electrolysis (PEMWE) as energy storage are its high current density, high purity gas production, and compact system. The membrane electrode assembly (MEA), including anode/electrolyte/cathode, is a key component of PEMWE. If the anode is composed of noble metal oxide-IrO_2, PEMWE produces hydrogen and oxygen at the cathode and anode side during the off-peak period for energy storage. The supply of stored hydrogen and oxygen gas for the fuel cell is used to generate power during the peak-hour period. If the anode of the MEA is composed of low-cost PbO_2 accompanied with a higher operating voltage, ozone gas will be generated in addition to oxygen. The PEMWE technique which uses MEA produces three kinds of gases (H_2, O_2, O_3) and will expand the application area [1–6].

In PEMWE, an anode with a proper catalyst enables a hydrogen production rate at a voltage of 2 V or less (Equation (1)). When seeking a higher production of hydrogen, the applied voltage is increased accompanied by a higher generated current as well as hydrogen (proton) according to the Faraday laws. Moreover, the anode catalyst is replaced by anti-corrosion material (ex. lead oxide) accompanied by a higher voltage applied, and ozone will be generated at the cathode in addition to oxygen (Equation (2)). Principles of PEMWE generating oxygen (ozone)/hydrogen and oxygen/hydrogen production are shown in the reaction equations are as follows:
Anode:

$$2H_2O \rightarrow O_2 + 4H^+ + 4e^- \quad (1.23\ V) \tag{1}$$

$$3H_2O \rightarrow O_2(O_3) + 6H^+ + 6e^- \quad (1.51\ V) \tag{2}$$

Cathode:

$$2H^+ + 2e^- \rightarrow H_2 \tag{3}$$

To prepare MEA for traditional proton exchange membrane fuel cells (PEMFC), spray catalyst ink on a gas diffusion layer (GDL) to form a gas diffusion electrode (GDE), and then stack the GDE of both cathode and anode with proton exchange membrane (as electrolyte) for hot pressing to form a complete three-layer MEA [7]. In addition, common methods to fabricate MEAs include the catalyst-coated membrane (CCM) process and the decal method. CCM means spraying catalyst inks directly on different sides of the membrane, and the decal method means coating catalyst inks on substrate to further transfer to both sides of the membrane. The three fabricated layers are then incorporated with two GDLs to form the complete MEA [8–14]. The MEA of PEMWE for hydrogen/oxygen production is mainly prepared [15–20] with the above methods. The CCM method is a small scale recommended for producing MEAs in the laboratory, and the decal method can be scaled-up as a pilot plant [21,22]. Common to these three methods described above is both cathode and anode catalysts are bonded strongly with the membrane and is named as membrane layer and abbreviated as ML. Use of ML to manufacture oxygen (ozone)/hydrogen and oxygen/hydrogen MEA is time-consuming and labor-intensive and can be a challenge to mass production. In this study, a new method called flat layer (FL) is proposed. The dried anode ink is flat-layer (FL) deposited and confined onto the membrane, whereas the cathode is hot pressed on the membrane the same as with ML. A characteristic of the FL method is to form a loose interface between the anode and membrane to avoid change during wet-operation or dry-closedown of PEMWE. FL, will not only be effective in significantly enhancing MEA production but it will also potentially improve the durability of PEMWE.

2. Materials and Methods

The electrolyte used for MEA production is DuPont's Nafion 117, a perfluorinated sulfonic acid membrane with a $40 \times 40\ mm^2$ area. The cathode gas diffusion electrode is from Yangtze Corporation (40% Pt/C catalyst, Pt loading $0.5\ mg/cm^2$). The electrolyzer housing is purchased from China Guangyuan Instrument (GQ-DJ800). The electrolyzer housing is to conduct both mass transfer as well as electric transfer of the MEA. The mass transfer includes injection of water to the anode while generating hydrogen and oxygen (or ozone) from the anode and cathode, respectively. The electric transfer includes an applied voltage to or generated current from the MEA. The assembly torque is $30\ kgf/cm^2$. The only difference lies in the anode preparation method for comparison. The ambient temperature is maintained at 20 °C via water cooling throughout the water electrolysis process.

2.1. Anode Preparation

For ML, coat the anode catalyst ink to the transfer substrate layer by layer and then dry it gradually. Once the dry anode catalyst ink on the transfer substrate reaches the set amount ($35\ mg/cm^2$), it is then transferred to the membrane with hot pressing (left of Figure 1). The reaction area is $30 \times 30\ mm^2$. After completion of the coating, the two-layer structure (anode/electrolyte) is stacked with the commercial cathode in a sandwich method for 2 min hot pressing at 135 °C resulting in a three-layer MEA. For the FL method, the commercial cathode and membrane are hot-pressed first to generate a two-layer structure. The prepared anode catalyst ink is dried at 90 °C for 50 min, then it is filled in a confined area onto the membrane side of the two-layer membrane/cathode structure (right of Figure 1). The amount of fill is equivalent to that of coating used in the previous ML. The time required for the two methods is very different. ML takes about 30 min to prepare an MEA, while FL only requires about 5 min to complete, so FL is superior in terms of production rate.

Figure 1. Anode side of (**a**) membrane-layer (ML) (left) and (**b**) flat-layer (FL) (right)

2.2. Electrolysis Test

Figure 2 shows the PEMWE test system. A constant-voltage of 4.5 V is applied to the electrolyzer (MEA + 2 End Plates), and the MEA performance is evaluated according to the generated current. It is operated by a power supplier (DR2002, MOTECH Corp., TAIWAN) which is shown in Figure 3. The higher the current, the more hydrogen and oxygen (ozone) is generated. In addition, according to the study noted by Onda et al., to produce hydrogen and oxygen (ozone) via PEMWE, MEA will reach a steady-state after 8 hours of activation, but the current will drop when power is interrupted and restored [23]. In this study, the MEA is powered by 4.5 V with a continuous 12-hour operation as activation. After the activation, the power is cut for 1 min, 10 min, and 1 h. After 1 min, 10 min, and 1 h, the MEA is restored at 4.5 V for 1 h to observe the current recovery (retaining the moisture of anode and cathode throughout the experiment).

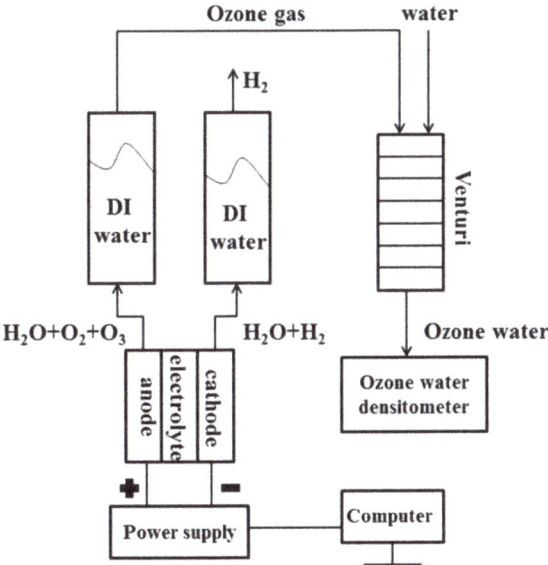

Figure 2. Proton exchange membrane water electrolysis (PEMWE) test system.

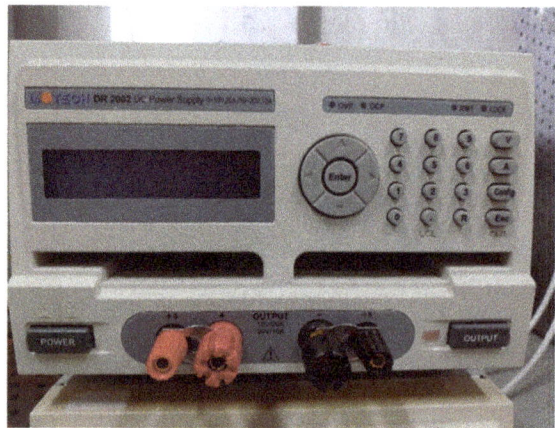

Figure 3. DR2002 power supplier.

2.3. Electrochemical Impedance Analysis

Use of electrochemical impedance spectroscopy to compare the MEA impedance of the two methods. The electrical resistance and charge transfer resistance of the MEAs fabricated with different methods are compared at high and low-frequency regions, respectively.

2.4. Accelerated Aging

2.4.1. High-Temperature Effect

Millet et al. reported the performance decline of electrolyzer caused by degradation while a high temperature will cause a perforated membrane and thus, a dangerous mix of hydrogen and oxygen [24]. Therefore, in this study after the 12 h MEA activation at 4.5 V, the MEA (as well as the end-plates) will be dried for 1 h at 150 °C (160 °C) and then powered for 10 min repeatedly so that the MEA (including membrane and the catalyst layer) are in harsh environments. The impacts of high temperature on the performance of the MEAs fabricated from the two methods are observed.

2.4.2. High-Voltage Effect

Theoretically, the voltages are 1.23 and 1.56 V for hydrogen/oxygen production and for hydrogen/oxygen(ozone) production, respectively [16]. In reality, the voltages needed are 2.0 V and 3.0 V for hydrogen/oxygen production and for hydrogen/oxygen(ozone) production, respectively. Although the higher the voltage and the higher the current, the more hydrogen and oxygen (ozone) is produced. The higher voltage and higher hydrogen/oxygen generated lead to harsh conditions encountered by the anode catalyst and both electrode structures. In this study, a higher voltage of 4.5~6.0 V is applied in order to learn the response of the MEAs fabricated from the two methods (ML and FL) through performance and durability tests.

3. Results and Discussion

A high voltage of 4.5 V is applied to the assembled electrolyzer and the ambient temperature is maintained at 20 °C during activation stages. High voltage is utilized in order to generate higher current as well as more generated oxygen/ozone gas. More generated oxygen/ozone gas will create harsh conditions to the anode structure as they penetrate through it. During the activation stages (0–12 h), the current of ML is slightly higher than that of FL, as shown in Figure 4. The currents of the two MEAs, ML and FL, at the end of the activation stages are 9.79 and 8.79 A, respectively. The performance of FL is 10.3% lower compared to that of ML. Both ML and FL exhibit a slow increase and stable state which proves the feasibility of FL. After power interruption and restoration, the currents of ML and FL decline by 30.5% and 22.4%, respectively. Similar trends had been reported by Onda et al. [23]. FL shows better resistance to lower performance compared to that of ML due to the dynamic state of the anode catalyst layer. A characteristic of the FL method is to form a loose interface between the anode and membrane to avoid change during wet-operation or dry-closedown of PEMWE. Therefore, FL can not only significantly enhance MEA production but also improve the durability of PEMWE after power interruption and restoration.

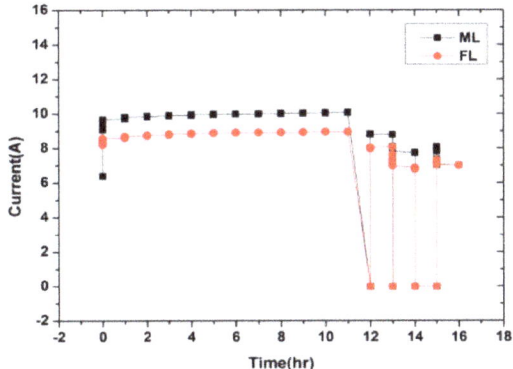

Figure 4. Accelerating test – high voltage with power interruption and restoration (4.5 V).

Figure 5 shows the analysis and comparison of the electrochemical impedance of ML and FL and the input voltage at 4.5 V. In the high-frequency portion, the impedance of ML is 0.18 Ω and that of FL is 0.27 Ω, indicating that the ohmic resistance of ML is less than that of FL. The lower the ohmic resistance, the higher the electrolytic current, which matches the current results from the activation stages as shown in Figure 4. In the low-frequency portion, the curve radius of ML (1.20 Ω) is less than that of FL (1.39 Ω), indicating that its catalytic reaction rate is better than that of FL [16]. Higher resistance of FL is due to its loose structure compared with ML, this will lead to lower conductivity resulting in a lower catalytic reaction rate.

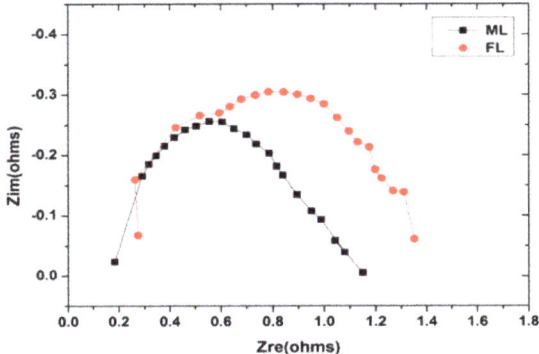

Figure 5. Electrochemical impedance analysis.

Under an accelerating test with high voltage (4.5 V) and high-temperature drying, both the currents of the MEAs of the two different methods after high-temperature and power restoration are much lower than the activation current as shown in Figure 6. Figure 6 shows the currents of the two methods decline by 81% (ML) and 90% (FL) when the power is restored after the first high temperature (drying), which indicates that neither of the two methods is able to work properly after high-temperature drying. After eight cycles, the current of ML declines by 90%, and that of FL declines by 97%, showing that in this operating environment, ML is slightly better than FL. The structure is fixed in the hot pressing process of ML at 135 °C, and only some flakes will peel off even after the high-temperature drying (left of Figure 7). The structure of FL after drying shows more cracks, as shown in the right of Figure 7. Furthermore, after disassembling the electrolyzer, we found that the structure of ML is fairly complete, while some areas of FL are white molten (right of Figure 7), which is due to the partial melting of polytetrafluoroethylene [18].

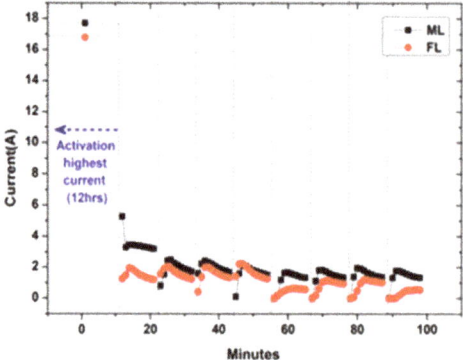

Figure 6. Accelerating electrolysis test—high voltage (4.5 V) and high-temperature drying (160 °C).

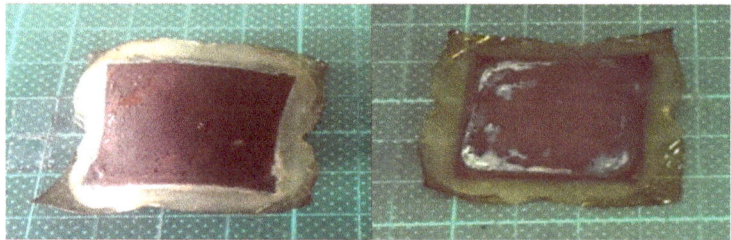

Figure 7. Anode structure after experiments (**a**) for ML and (**b**) for FL.

The applied voltage is further increased from 4.5, 5.5, to 6 V, and the generated current of ML and FL after 12 h activation are shown in Figure 8. When the voltage is 4.5 V, the average current of FL is approximately 10.2% lower than that of FL. As it increases to 5.5 V and 6.0V, the difference is 2.8% and 1.3%, respectively. The results indicate that in the case of high voltage, the performance of FL is lower than that of ML. With the voltage further increased and thus, the deterioration of the environment, FL narrows the gap with ML. The performance of electrolysis increases with voltage, indicating the increased anode ozone and oxygen production and increased overall gas output will cause more structural damage to ML compared to that of FL. The results show that in the case of ultra-high voltage and large anode gas production, the process of FL will be better than that of ML.

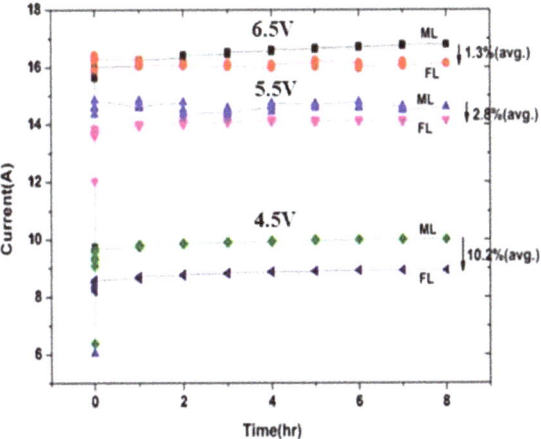

Figure 8. Current differences under different applied voltages.

According to the experimental results from Figures 4–8, structural simulation of both ML and FL after the accelerating test (high-temperature drying + high-voltage applied) is shown in Figure 9. Before the accelerating test, the anode structure (PbO_2) of ML (bottom left) is arranged closely to the Nafion membrane (electrolyte) compared to loosely contacted in that of FL (bottom right). During the accelerating test (high voltage applied + high-temperature drying), the interface between anode and Nafion is weakened during high-temperature drying and wet operation, whereas some PbO_2 particles are carried away from their position due to high O_2/O_3 flow rate (center left). If the temperature and voltage are further increased, the anode structure will be more greatly deteriorated (top left). Due to the loose contact between anode and Nafion membrane (electrolyte) accompanied by a weak connection between anode particles (PbO_2), the effect of high-temperature drying and high-voltage applied on the change of the anode structure is minor (center right, top right) compared to that ML.

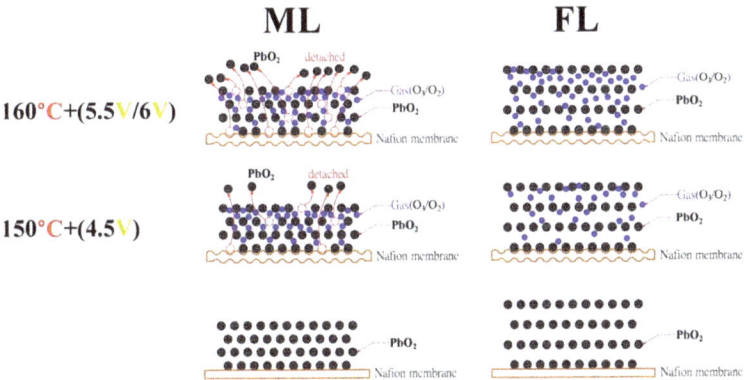

Figure 9. Structural simulation after accelerating test for ML (left) and for FL (right).

4. Conclusions

MEA preparation in the PEMWE technology is time consuming and labour-intensive, and the experimental test will damage the anode structure of the traditional ML and thus, its performance. The innovative FL thereby derived is proven to have a better preparation rate than ML in practical tests and can effectively save more than 80% labor time. The anode catalyst layer of FL retains a loose and unstable/flowable structure, thus its conductivity is lower compared to the solid and stable structure as in the traditional ML. Therefore, for operation of electrolyzer under normal voltage or temperature, performance/durability of ML is better than that of FL. For operation of electrolyzer under a higher voltage or temperature for more production of hydrogen and oxygen (ozone), the solid and stable structure of ML would be a resistance for more generated hydrogen and oxygen (ozone) to penetrate. However, the higher the voltage and the harsher the operation conditions, the current of FL gradually approaches that of ML.

In summary, FL is slightly inferior to ML in terms of electrolytic effect in the general environment, but in the face of harsher operation conditions, the performance of FL gradually approaches or is more superior than that of ML, which also provides a better alternative for medical purposes and sterilization that require high oxygen and ozone concentrations.

Author Contributions: Conceptualization, G.-B.J. and C.-C.Y.; Methodology, G.-B.J. and S.-H.C.; Validation, G.-B.J. and S.-H.C.; Formal Analysis, G.-B.J., C.-C.Y. and C.-J.L.; Investigation, G.-B.J., C.-C.Y. and J.-W.Y.; Data Curation, C.-J.L.; Writing-Original Draft Preparation, J.-W.Y. and G.-B.J.; Writing-Review & Editing, J.-W.Y. and G.-B.J.; Visualization, J.-W.Y.; Supervision, Shih-Hung Chan; Project Administration, S.-H.C.; Funding Acquisition, G.-B.J.

Funding: The authors are grateful to the Ministry of Science and Technology of Taiwan, under contracts MOST106-2221-E-155-042, MOST108-3116-F-155-001 and MOST108-2221-E-155-002 for their financial support.

Conflicts of Interest: The authors declare no conflict of interest.

References

1. Buttler, A.; Spliethoff, H. Current status of water electrolysis for energy storage, grid balancing and sector coupling via power-to-gas and power-to-liquids: A review. *Renew. Sustain. Energy Rev.* **2018**, *82*, 2440–2454. [CrossRef]
2. Fallisch, A.; Schellhase, L.; Fresko, J.; Zechmeister, M.; Zedda, M.; Ohlmann, J.; Zielke, L.; Paust, N.; Smolinka, T. Investigation on pem water electrolysis cell design and components for a hycon solar hydrogen generator. *Int. J. Hydrogen Energy* **2017**, *42*, 13544–13553. [CrossRef]
3. Kopp, M.; Coleman, D.; Scheppat, B.; Stiller, C.; Scheffer, K.; Aichinger, J. Energiepark mainz: Technical and economic analysis of the worldwide largest power-to-gas plant with pem electrolysis. *Int. J. Hydrogen Energy* **2017**, *42*, 13311–13320. [CrossRef]
4. Barbir, F. Pem electrolysis for production of hydrogen from renewable energy sources. *Sol. Energy* **2005**, *78*, 661–669. [CrossRef]
5. Carmo, M.; Fritz, D.L.; Mergel, J.; Stolten, D. A comprehensive review on pem water electrolysis. *Int. J. Hydrogen Energy* **2013**, *38*, 4901–4934. [CrossRef]
6. Chen, S.; Jiang, F.; Xie, X.; Zhou, Y.; Hu, X. Synthesis and application of lead dioxide nanowires for a pem ozone generator. *Electrochim. Acta* **2016**, *192*, 357–362. [CrossRef]
7. Bussayajarn, N.; Ming, H.; Hoong, K.K.; Ming Stephen, W.Y.; Hwa, C.S. Planar air breathing pemfc with self-humidifying mea and open cathode geometry design for portable applications. *Int. J. Hydrogen Energy* **2009**, *34*, 7761–7767. [CrossRef]
8. Hu, M.; Sui, S.; Zhu, X.; Yu, Q.; Cao, G.; Hong, X.; Tu, H. A 10 kw class pem fuel cell stack based on the catalyst-coated membrane (ccm) method. *Int. J. Hydrogen Energy* **2006**, *31*, 1010–1018. [CrossRef]
9. Chun, J.H.; Park, K.T.; Park, S.H.; Jo, D.H.; Lee, E.S.; Lee, J.Y.; Kim, S.G.; Kim, S.H.; Jyoung, J.-Y. Development of a novel hydrophobic/hydrophilic double micro porous layer for use in a cathode gas diffusion layer in pemfc. *Int. J. Hydrogen Energy* **2011**, *36*, 8422–8428. [CrossRef]
10. Liu, C.-H.; Ko, T.-H.; Shen, J.-W.; Chang, S.-I.; Chang, S.-I.; Liao, Y.-K. Effect of hydrophobic gas diffusion layers on the performance of the polymer exchange membrane fuel cell. *J. Power Sources* **2009**, *191*, 489–494. [CrossRef]
11. Zhao, X.; Fu, Y.; Li, W.; Manthiram, A. Effect of non-active area on the performance of subgasketed meas in pemfc. *Int. J. Hydrogen Energy* **2013**, *38*, 7400–7406. [CrossRef]
12. Kim, K.-H.; Kim, H.-J.; Lee, K.-Y.; Lee, S.-Y.; Cho, E.; Lim, T.-H.; Yoon, S.P.; Hwang, I.C.; Jang, J.H. The effects of nafion® ionomer content in pemfc meas prepared by a catalyst-coated membrane (ccm) spraying method. *Int. J. Hydrogen Energy* **2010**, *35*, 2119–2126. [CrossRef]

13. Kim, K.-H.; Lee, K.-Y.; Lee, S.-Y.; Cho, E.; Lim, T.-H.; Kim, H.-J.; Yoon, S.P.; Kim, S.H.; Lim, T.W.; Jang, J.H. The effects of relative humidity on the performances of pemfc meas with various nafion® ionomer contents. *Int. J. Hydrogen Energy* **2010**, *35*, 13104–13110. [CrossRef]
14. Yang, T.-F.; Cheng, C.-H.; Su, A.; Yu, T.-L.; Hourng, L.-W. Numerical analysis of the manipulated high performance catalyst layer design for polymer electrolyte membrane fuel cell. *Int. J. Energy Res.* **2014**, *38*, 1937–1948. [CrossRef]
15. Xu, W.; Scott, K. The effects of ionomer content on pem water electrolyser membrane electrode assembly performance. *Int. J. Hydrogen Energy* **2010**, *35*, 12029–12037. [CrossRef]
16. Song, S.; Zhang, H.; Ma, X.; Shao, Z.; Baker, R.T.; Yi, B. Electrochemical investigation of electrocatalysts for the oxygen evolution reaction in pem water electrolyzers. *Int. J. Hydrogen Energy* **2008**, *33*, 4955–4961. [CrossRef]
17. Su, H.; Linkov, V.; Bladergroen, B.J. Membrane electrode assemblies with low noble metal loadings for hydrogen production from solid polymer electrolyte water electrolysis. *Int. J. Hydrogen Energy* **2013**, *38*, 9601–9608. [CrossRef]
18. Xu, J.; Miao, R.; Zhao, T.; Wu, J.; Wang, X. A novel catalyst layer with hydrophilic–hydrophobic meshwork and pore structure for solid polymer electrolyte water electrolysis. *Electrochem. Commun.* **2011**, *13*, 437–479. [CrossRef]
19. Rozain, C.; Mayousse, E.; Guillet, N.; Millet, P. Influence of iridium oxide loadings on the performance of pem water electrolysis cells: Part I—Pure iro2-based anodes. *Appl. Catal. B Environ.* **2016**, *182*, 153–160. [CrossRef]
20. Yu, J.-W.; Jung, G.-B.; Su, Y.-J.; Yeh, C.-C.; Kan, M.-Y.; Lee, C.-Y.; Lai, C.-J. Proton exchange membrane water electrolysis system-membrane electrode assembly with additive. *Int. J. Hydrogen Energy* **2019**, *44*, 15721–15726. [CrossRef]
21. Burdzik, A.; Stähler, M.; Friedrich, I.; Carmo, M.; Stolten, D. Homogeneity analysis of square meter-sized electrodes for pem electrolysis and pem fuel cells. *J. Coat. Technol. Res.* **2018**, *15*, 1423–1432. [CrossRef]
22. Mauger, S.A.; Neyerlin, K.C.; Yang-Neyerlin, A.C.; More, K.L.; Ulsh, M. Gravure coating for roll-to-roll manufacturing of proton-exchange-membrane fuel cell catalyst layers. *J. Electrochem. Soc.* **2018**, *165*, 1012–1018. [CrossRef]
23. Onda, K.; Ohba, T.; Kusunoki, H.; Takezawa, S.; Sunakawa, D.; Araki, T. Improving characteristics of ozone water production with multilayer electrodes and operating conditions in a polymer electrolyte water electrolysis cell. *J. Electrochem. Soc.* **2005**, *152*, 177–183. [CrossRef]
24. Millet, P.; Ranjbari, A.; De Guglielmo, F.; Grigoriev, S.A.; Auprêtre, F. Cell failure mechanisms in pem water electrolyzers. *Int. J. Hydrogen Energy* **2012**, *37*, 17478–17487. [CrossRef]

© 2019 by the authors. Licensee MDPI, Basel, Switzerland. This article is an open access article distributed under the terms and conditions of the Creative Commons Attribution (CC BY) license (http://creativecommons.org/licenses/by/4.0/).

Article

Effect of Dispersion Solvents in Catalyst Inks on the Performance and Durability of Catalyst Layers in Proton Exchange Membrane Fuel Cells

Chan-Ho Song and Jin-Soo Park *

Department of Green Chemical Engineering, College of Engineering, Sangmyung University,
C-411 Main Building, 31 Sangmyungdae-gil, Dongnam-gu, Cheonan 31066, Korea; schanho0525@gmail.com
* Correspondence: energy@smu.ac.kr; Tel.: +82-41-550-5315

Received: 24 January 2019; Accepted: 8 February 2019; Published: 11 February 2019

Abstract: Five different ionomer dispersions using water–isopropanol (IPA) and N-methylpyrrolidone (NMP) were investigated as ionomer binders for catalyst layers in proton exchange membrane fuel cells. The distribution of ionomer plays an important role in the design of high-performance porous electrode catalyst layers since the transport of species, such as oxygen and protons, is controlled by the thickness of the ionomer on the catalyst surface and the continuity of the ionomer and gas networks in the catalyst layer, with the transport of electrons being related to the continuity of the carbon particle network. In this study, the effect of solvents in ionomer dispersions on the performance and durability of catalyst layers (CLs) is investigated. Five different types of catalyst inks were used: (i) ionomer dispersed in NMP; (ii) ionomer dispersed in water–IPA; (iii) ionomer dispersed in NMP, followed by adding water–IPA; (iv) ionomer dispersed in water–IPA, followed by adding NMP; and (v) a mixture of ionomer dispersed in NMP and ionomer dispersed in water–IPA. Dynamic light scattering of the five dispersions showed different average particles sizes: ~0.40 µm for NMP, 0.91–1.75 µm for the mixture, and ~2.02 µm for water–IPA. The membrane-electrode assembly prepared from an ionomer dispersion with a larger particle size (i.e., water–IPA) showed better performance, while that prepared from a dispersion with a smaller particle size (i.e., NMP) showed better durability.

Keywords: catalyst layer; ionomer dispersion; durability; proton exchange membrane fuel cell

1. Introduction

Enhancement of the performance and durability of proton exchange membrane fuel cells (PEMFCs), finally resulting in cost reduction in fuel cells, is essential for their commercialization [1–3]. The performance and durability of PEMFCs are typically determined by the polymer electrolyte membrane (also called an ionomer membrane), which are mostly made of perfluorinated sulfonic acid (PFSA) and porous electrodes comprising electrocatalysts and an ionomer binder in membrane-electrode assemblies (MEAs) [3–6]. Optimization of porous electrodes has been achieved by controlling relative amounts of the ionomer and/or electrocatalyst if the polymer electrolyte membranes are fixed [7–10]. It has been reported that the distribution of ionomer within the carbon supported electrocatalyst (e.g., Pt/C) plays an important role in the design of high-performance porous electrode catalyst layers (CLs), as the transport of species such as oxygen and protons is controlled by the thickness of the ionomer on the catalyst surface and the continuity of the ionomer and gas networks in the CL, with the transport of electrons being related to the continuity of the carbon particle network [11–16]. The structural change of CLs significantly influences Pt utilization by the attachment of ionomers onto the carbon support of the Pt electrocatalyst [17]. The use of new ionomer materials to enhance the transport phenomena occurring in CLs could be another way to increase performance and/or durability. Some ionomer materials represent a high oxygen permeability to lower the mass transfer resistance for an oxygen reduction reaction by changing chemical structure of ionomers [18,19], or represent high proton conductivity by lowering equivalent weight of ionomers [20]. The microstructure of a CL is formed by evaporating solvents included in the catalyst ink coated onto polymer electrolyte membranes or gas diffusion layers. The catalyst ink is a well-stirred mixture of an ionomer binder dispersion, carbon-supported electrocatalysts, and/or additional solvents. In the ionomer binder, Nafion is dispersed as particles with three types of morphology: (i) a well-defined cylindrical dispersion in glycerol and in ethylene glycol with different degrees of solvent penetration; (ii) a less-defined, highly-solvated large particle in water–isopropanol (water–IPA) mixtures; and (iii) a random-coil conformation (true solution behavior) in N-methylpyrrolidone (NMP) [21]. The various shapes of Nafion dispersed in these three types of solvent could influence the performance and durability of the PEMFCs. Johnston and co-workers reported that cathodes cast from inks based on ionomer dispersions in water–propanol–isopropanol initially performed better than those cast from glycerol-based dispersions, but were far less durable because of a higher degree of phase separation, which resulted in faster Pt particle growth [22]. The types of solvents used affected the surface morphology of the CLs. Inks containing high-boiling-point solvents (186–212 °C), such as glycerin, ethylene glycol, and propylene glycols, resulted in CLs with fewer cracks than those made with inks containing low-boiling-point solvents (65–100 °C), such as water or methanol [23]. In spite of the superior durability of the CLs based on a high-boiling-point solvent, glycerol-processed cathode CLs showed substantially inferior performance than NMP-processed cathode CLs, which are similar to water–IPA-processed ones [24].

Herein, the effect of solvents in ionomer dispersions on the performance and durability of CLs is investigated. Five different types of catalyst inks were used: (i) ionomer dispersed in NMP (coded as "2.5 wt.% NMP"); (ii) ionomer dispersed in water–IPA (coded as "2.5 wt.% water–IPA"); (iii) ionomer dispersed in NMP, followed by adding water–IPA (coded as "5 wt.% NMP+adding water–IPA"); (iv) ionomer dispersed in water–IPA, followed by adding NMP (coded as "2.5 wt.% water–IPA+adding NMP"); and (v) a mixture of ionomer dispersed in NMP and ionomer dispersed in water–IPA (coded as "2.5 wt.% water–IPA+2.5 wt.% NMP"). The size distributions of ionomers dispersed in the five different solvent systems were measured by dynamic light scattering (DLS). For the characterization of the CLs, five different membrane-electrode assemblies were fabricated and characterized by physical methods, such as scanning electron microscopy (SEM), and electrochemical methods, such as cyclic voltammetry and current-voltage polarization.

2. Materials and Methods

Five different ionomer dispersions with a concentration of 2.5 wt.% were prepared by dispersing tiny pieces of Nafion NR-212 membrane (Dupont) in the corresponding solvents, such as a 50:50 (wt.%) water–IPA mixture, NMP, and mixtures of water–IPA and NMP. The ionomer dispersions based on NMP and water–IPA were heated stepwise to 70, 80, and 90 °C under vigorous magnetic stirring, in glass vials sealed with gas impermeable covers to prevent solvent evaporation. The solutions were held at each temperature for 3 h, and the final temperature was maintained until the dispersion became transparent. The addition of the secondary solvent for the ionomer dispersions based on NMP followed by adding water–IPA, and the dispersions based on water–IPA followed by adding NMP, was carried out at room temperature (20 ± 1 °C), and the mixtures were kept stirring for 1 day. Dynamic light scattering (ELSZ–1000, Photal Otsuka Electronics Co., Ltd., Osaka, Japan) was performed to measure the ionomer size distributions in the dispersions. Membranes with 50 μm in thickness were cast from the ionomer dispersions for the measurement of proton conductivity. The ionomer dispersions were poured into glass petri dishes and dried at 80 °C under vacuum for 2 days to allow complete evaporation of solvents. The in-plane proton conductivity of the membranes was determined by measuring the impedance in a plate cell with four Pt wire electrodes. The impedance for ion conductivity of fully wet membrane samples equilibrated in 0.1 M H_2SO_4 for 1 day was measured using a potentiostat (SP-150, Bio-Logic Science Instruments, Seyssinet-Pariset, France) with alternating current (AC) in the scan range of 1 MHz to 1 Hz, with a signal amplitude of 10 mV for 1 min at 20 ± 1 °C. The impedance was obtained at zero phase angle. For catalyst ink preparation, 40 wt.% Pt was deposited on Vulcan XC-72 carbon black (Hispec 4000, Johnson-Matthey, Pennsylvania, PA, USA) and one of the lab-made dispersions was prepared by magnetic stirring and sonication for 30 min. Magnetic stirring and sonication were repeated 4 times. The catalyst inks were cast on polyimide transfer films by a Meyer bar installed in a Meyer bar coater (KP-3000, KIPAE E&T Co., Inc., Suwon, Korea), and the catalyst-coated films were dried at 80 °C for 48 h to allow complete evaporation of solvents. The coated catalyst layers with an area of 9 cm^2 (3 × 3 cm^2) were transferred onto the surface of Nafion 211 membranes under 30 MPa for 3 min. The final Pt loading amount was 0.4 mg/cm. Surface images of the catalyst-coated membranes (CCMs) were obtained by scanning electron microscopy (MIRA LMH, TESCAN, Brno, Czech Republic). The CCMs were sandwiched by a pair of gas diffusion layers (10BC, SIGRACET®, SGL Group–The Carbon Company, Wiesbaden, Germany). The MEAs were placed in the unit cell with a pair of bipolar plates with a single serpentine channel and were then assembled at 50 kgf/cm^2. The MEAs were evaluated using a fuel cell station (CNL Energy Co., Inc., Seoul, Korea) connected to a potentiostat (SP-150, Bio Logic Science Instruments, Seyssinet-Pariset, France) in terms of current-voltage polarization and cyclic voltammetry. The cell temperature was kept at 70 °C, and hydrogen and air humidified to 100% relative humidity (RH) were supplied to the anode and cathode, respectively, at ambient pressure. After the preset temperatures were achieved, hydrogen and oxygen were supplied to the cell at a rate of 200 ccm (cc/min). The MEAs were conditioned only 10 times by repeatedly running a voltage cycle from an open circuit voltage (OCV) to the voltage corresponding to the maximum current density to investigate the effect of the solvent system used in the ionomer dispersions on unsteady-state phenomena (e.g., reorientation of ionomer in CLs) of CLs based on various ionomer dispersions. After the conditioning, the current-voltage data were logged. The electrochemically active surface area (ECSA) was obtained from the cyclic voltammograms of MEAs as explained elsewhere [3]. To test the durability, the electrodes were subjected to 5, 10, 15, 20, 25, and 30k potential cycles (k means ×1000) at 50 mV/s from 0.60 to 1.0 V, based on the U.S. Department of Energy (DOE) protocol for testing electrocatalysts [25].

3. Results and Discussion

Recent studies, reporting that the solvent type significantly affected the shape of the ionomer dispersed in the solvents, pointed out that polar protic solvents, such as water, *n*-propanol, isopropanol, ethanol, or methanol, result in highly-solvated large particles, while polar aprotic solvents, such as NMP, dimethylformamide, or dimethyl sulfoxide, result in a random-coil conformation (true solution behavior) [21–24]. Table 1 summarizes the size distributions of the five different ionomer dispersions measured by DLS. The average ionomer size for the water–IPA dispersion was the highest among the five dispersions, and the smallest size was observed for the NMP dispersion, which is in good agreement with the previous study [21]. Interestingly, the dispersions prepared by the solvent mixture including water–IPA and NMP had intermediate average ionomer sizes, as compared to the one prepared using either water–IPA or NMP only. It means that ionomers solvated by water–IPA and NMP are mixed homogeneously. For dispersions in which the additional solvents were added, the additional solvents substantially determined the ionomer particle structure. The addition of NMP in 5 wt.% water–IPA and water–IPA in 5 wt.% NMP (final concentration is 2.5 wt.%) resulted in the dispersion similar to the 2.5 wt.% NMP dispersion and the dispersion similar to the 2.5 wt.% water–IPA based dispersion, respectively.

Figure 1 shows proton conductivity of the five different cast membranes and Nafion NR-212 membranes as a reference. Since Nafion NR-212 is also formed in a dispersion casting process, most of the five different cast membranes show similar or a little higher proton conductivity as compared to that of Nafion NR-212. It represents a change in the ionomer properties in different solvent systems due to the different degree of phase separation of the PFSA ionomers. It does not, of course, mean that the proton conductivity of the various ionomers proportionally represents MEA performance, as the properties of the films as membrane electrolytes prepared by ionomers in various solvent systems and the ultra-thin film as ionomer binder (normally wrapping electrocatalyst particles) are not the same as those reported in literature [26]. It is noted that the change of solvent systems for ionomer dispersions could significantly affect ionomer properties. The structural changes of ionomers in solvents could cause hydrophobic and hydrophilic phase separation by the backbone and side chains of perfluorinated sulfonic acid (PFSA). To confirm the change in phase separation behavior, membranes were cast from the five different dispersions and their proton conductivities were measured in-plane (Figure 1). The proton conductivities of the membranes should be the same since the same ionomer material was used for the five different ionomer dispersions. However, proton conductivities in the range 0.75–0.83 S/cm (77%–86% of the proton conductivity for the water–IPA ionomer dispersion) were obtained for the membranes cast from ionomer dispersions including NMP.

Table 1. Summary of preparation methods of five different ionomer dispersions and their dynamic light scattering (DLS).

2.5 wt.% Dispersions		Average Diameter (μm)	D10 [1]	D50 [1]	D90 [1]	Span [2]
Main Dispersion	Additional Solvent					
2.5 wt.% water–IPA (1:1)	-	2.00 ± 0.03	0.440 ± 0.106	34.7 ± 7.7	62.4 ± 3.5	1.78
2.5 wt.% NMP	-	0.400 ± 0.023	0.028 ± 0.023	0.297 ± 0.072	0.849 ± 0.167	2.76
2.5 wt.% water–IPA (1:1)+2.5 wt.% NMP	-	0.905 ± 0.017	0.094 ± 0.075	1.32 ± 0.073	6.60 ± 0.77	4.95
5 wt.% water–IPA (1:1)	NMP	1.20 ± 0.04	0.046 ± 0.017	2.11 ± 0.62	7.32 ± 1.80	3.45
5 wt.% NMP	Water–IPA (1:1)	1.75 ± 0.07	0.20 ± 0.04	3.61 ± 0.21	12.8 ± 2.81	3.50

[1] Dxx means that xx% of the ionomer particles have a size of the corresponding figure or smaller. [2] Span (=(D90−D10)/D50) indicates how far apart the 10% and 90% points are.

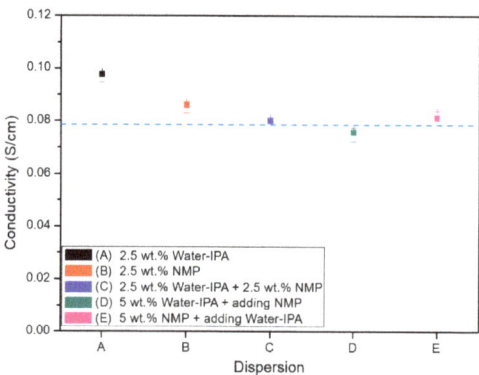

Figure 1. In-plane proton conductivity of the membranes cast from the five different ionomer dispersions prepared in this study (the dotted line represents the in-plane proton conductivity of Nafion 212).

To investigate the effect of the solvent system used for ionomer dispersion on the performance of CLs in MEAs, the electrochemical surface area via cyclic voltammetry and current-voltage polarization was measured. Figure 2 shows the variation of ECSA with respect to the number of potential cycles. All the MEAs based on the five different ionomer dispersions showed a tendency to decrease ECSA, which is the area of electrochemically reactive surface sites per amount of Pt (called three-phase boundary, TPB) and is obtained by recording the total charge required for the adsorption–desorption of a monolayer of hydrogen onto the Pt/ionomer interface as the number of potential cycles is increased. The ECSA of the MEA based on NMP ionomer dispersion; however, increased a little bit at the beginning of the potential cycling (0–10k) and then decreased gradually. It could; therefore, be inferred that during potential cycling, the ionomer, formed by the dispersion based on NMP, in the CLs of the MEA experiences reorientation resulting in structural change of TPB. It could suggest that MEAs with the CLs based on NMP ionomer dispersions require full conditioning to achieve steady-state performance.

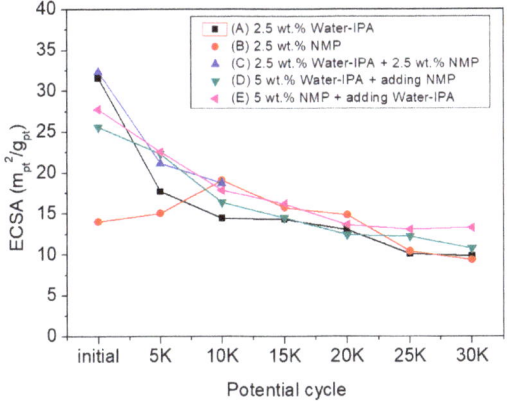

Figure 2. ECSA of MEAs prepared from the five dispersions at 0.6 V as a function of the number of potential cycles.

In Figure 3, the MEA based on NMP shows the lowest performance, but obtained the highest performance increment of 226% compared to the initial performance. In addition, the MEAs based on the ionomer dispersion in the mixed solvent system including NMP showed a modest increase in performance and then decreased slightly up to performance variations of −3.76%, 14.3%, and 26.4% at 30k potential cycles, compared to their initial performance for the MEAs based on 2.5 wt.% water–IPA+2.5 wt.% NMP; 5 wt.% water–IPA+adding NMP; and 5 wt.% NMP+adding water–IPA, respectively (positive and negative indicate increment and decrement, respectively). The numerical values of relative performance variation showed some fluctuation, but the performances remained mostly constant throughout the potential cycling. It could; therefore, be concluded that ionomers in the CLs based on ionomer dispersions in solvent systems including NMP need to experience reorientation finally resulting in the structural change of TPB. In addition, longer conditioning of these MEAs is needed to obtain steady-state operation. This could be because NMP results in a random-coil conformation (true solution behavior) in contrast to the other solvents (i.e., a well-defined cylindrical dispersion in glycerol and in ethylene glycol and a less-defined, highly solvated large particle morphology in water–IPA mixtures). This random coil formation results in unsteady-state hydrophilic-hydrophobic segregation of ionomers and makes the CLs require more time to achieve a steady-state TPB with the electrocatalyst and ionomer, due to the lower degree of phase-separation of the ionomer formed by the 2.5 wt.% NMP ionomer dispersion. The lower degree of phase-separation in the electrode retards a decrease in ECSA (in other words, an increase in Pt particle growth) in the CLs as shown in Figure 2, and finally results in a more durable TPB, in good agreement with previous results [23]. It means that the structure of TPB could be significantly determined by the shape of ionomers dispersed in solvents, not a specific property of solvents. It was reported that the shape of ionomers affects platinum particle growth due to the different mobility of side or main chains to the solvents [21]. In addition, penetration of the bulky ionomer dispersed in water–IPA to micropores containing catalyst in CLs appears implausible. For mesopores of CLs formed by mesoporous carbon support structures, as much as 50% of the Pt catalyst were not in direct contact with ionomer [11,27]. Thus, it could be expected that tiny ionomer from NMP-based dispersions wrap up electrocatalyst particles well to form a more compact TPB structure. It also influences platinum growth to determine the performance and/or durability rather than TPB structures formed by water–IPA dispersions. In summary, three I–V curves of the MEAs, with CLs prepared by the mixture of water–IPA and NMP (see Figure 3c–e), shows similar behavior in the region of activation, Ohmic and mass transport polarization. Significant difference in the region of activation and mass transport polarization was; however, shown in Figure 3a,b. The lower degree of phase-separation in the CL by the NMP dispersion caused higher activation polarization than the CL by the water–IPA dispersion. Since the relative dense CL by the NMP dispersion had smaller size pores where flooding occurred, at RH 100%, the mass transport polarization was less sensitive than the relative sparse CL by the water–IPA dispersion. Figure 4 summarizes the current density at 0.6 V obtained from all the current-voltage polarization curves of the five different MEAs, as a function of the number of potential cycles. Degradation of MEAs based on ionomer dispersions, using water–IPA with repeated fuel cell runs, resulted in a performance variation of −28.1% after 30 k potential cycles, compared to the initial performance due to a decrease in ECSA by Oswald ripening of Pt particles, oxidation of carbon supports, and/or cleavage of carbon-sulfur bonds in PFSA ionomers [28]. As discussed in Figure 2, the reorientation of ionomers in the CLs, formed by the ionomer dispersion based on NMP, changed the structure of the TPB of the electrodes and then resulted in an increase in ECSA. In addition, as reported in elsewhere [24], the surface of the initial CL, based on ionomer dispersions including NMP, showed no cracks, as shown in Figure 5b–e, but numerous cracks were observed for the CL based on 2.5 wt.% water–IPA (1:1), shown in Figure 5a. In other words, the dispersions containing a high-boiling-point solvents, such as NMP (see Figure 5b–e), showed less cracks than the dispersion containing low-boiling-point solvents, such as water and IPA (see Figure 5a). The solvents with low boiling points (less than ~100 °C) evaporated much faster than that with high boiling points (greater than ~150 °C). In addition, the clustering of highly-solvated large particles by water–IPA was suddenly vanished by fast evaporation. Thus, big voids could be formed in CLs. Consequently, voids inside CLs caused numerous cracks on the surface of CLs, as shown in Figure 5a. It is also the main reason for the better durability of the MEAs with CLs based on ionomer dispersions, including NMP. Nevertheless, the better performance of the MEA based on water–IPA is slightly attributed to the surface cracks since the recent study reported that cracked CLs by mechanical stretching of catalyst-coated Nafion membranes led to a decrease in membrane resistance and an improvement in mass transport, which resulted in enhanced device performance [29].

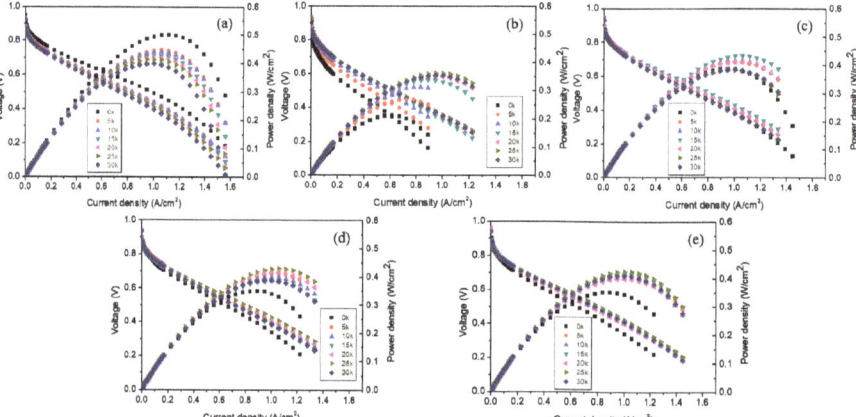

Figure 3. Fuel cell polarization plots after 5, 10, 15, 20, 25, and 30k potential cycles: (**a**) 2.5 wt.% water–IPA; (**b**) 2.5 wt.% NMP; (**c**) 2.5 wt.% water–IPA+2.5 wt.% NMP; (**d**) 5 wt.% water–IPA+adding NMP; and (**e**) 5 wt.% NMP+adding water–IPA.

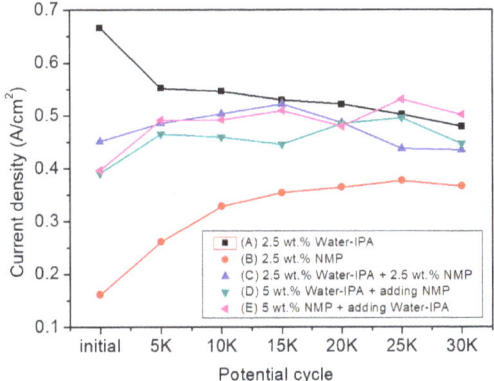

Figure 4. Current density of MEAs prepared from the five dispersions at 0.6 V as function of the number of potential cycles.

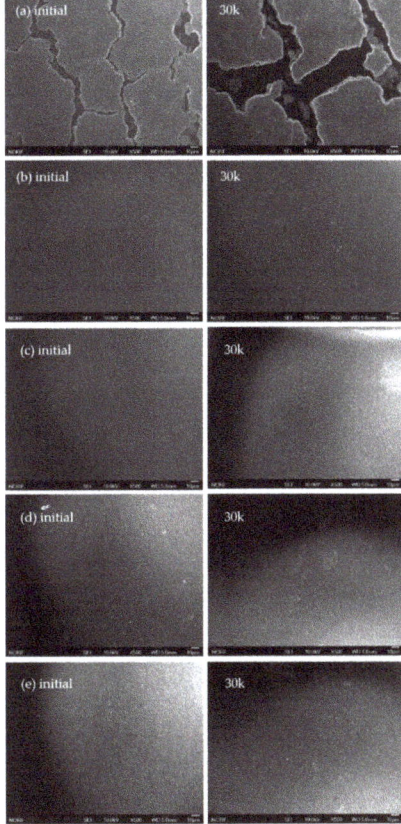

Figure 5. Scanning electron microscope (SEM) images of the surface of CCMs prepared from: (**a**) 2.5 wt.% water–IPA; (**b**) 2.5 wt.% NMP; (**c**) 2.5 wt.% water–IPA+2.5 wt.% NMP; (**d**) 5 wt.% water–IPA+adding NMP; and (**e**) 5 wt.% NMP+adding water–IPA dispersions, both initially and after 30k potential cycles at 500× magnification.

Figure 6 summarizes the performance (i.e., maximum current density during potential cycles and current density at 0.6 V and the 30kth potential cycle) and durability (i.e., two regressed lines showing that the discrepancy between two lines enlarges as the size of ionomers increases) of the MEAs with CLs based on the five different ionomer dispersions as a function of average particle size of the ionomer dispersions. Water–IPA results in highly-solvated large particles, with approximate diameters of >2 μm, and NMP results in a random coil-conformation, with approximate diameter of ~0.4 μm. The water–IPA based MEA showed the highest performance, and the NMP based MEA showed the lowest performance during the whole potential cycling test. However, the durability showed the totally opposite behavior. The final performances at the 30kth potential cycle were compared to maximum performances during potential cycles. It is observed that the discrepancy between two regressed lines decreases as the size of ionomer particles in dispersions decreases. Small discrepancy means small degradation of CLs during potential cycles. Hence, the ionomer dispersions based on mixtures of water–IPA and NMP show intermediate performance and durability. On the basis of these results, a relationship between performance-durability and average particle size in ionomer dispersions was found, as shown in Figure 6. That is, better performance could be obtained for MEAs with CLs based on ionomer dispersions with high average particle size, and better durability with lower average particle size. Once NMP participates in the solvent system of ionomer dispersions, the average ionomer particle size in the ionomer dispersions decreased below that of a water–IPA based ionomer dispersion. In other words, MEAs with CLs based on ionomer dispersions including NMP could exhibit lower performance and higher durability than those based on a water–IPA ionomer dispersion. The average particle size of ionomer dispersions could be an important indicator to predict the performance and durability of MEAs in PEMFCs.

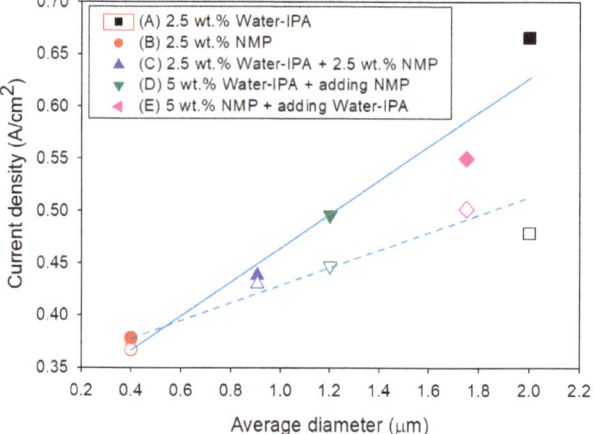

Figure 6. Current density of the MEAs at 0.6 V as function of the average size of ionomers in five different dispersions (the solid and blank symbols represent the maximum current densities during potential cycles and the current densities at the 30kth potential cycle, respectively; and the R^2 and p-value of the regressed solid line and the regressed dotted line are 0.97 and 0.007 and 0.80 and 0.04, respectively).

4. Conclusions

CLs using the dispersions prepared by five different solvents were investigated to understand the substantial effect of solvents on the morphology of ionomers dispersed in the various solvent systems, on the structure of the electrode surface, and on TPB in CLs. In addition, the effect of solvents, used in ionomer dispersions, on the performance and durability of MEAs was investigated. The ionomer dispersions based on five different solvent systems using water–IPA and/or NMP was used for investigation. Polar protic solvents, such as water and alcohols, resulted in good performance but inferior durability, due to the lower degree of phase-separation and numerous cracks on the surface of CLs, while polar aprotic solvents, such as NMP, showed more durable performance. Thus, ionomer dispersions in the solvent mixture of water–IPA and NMP were mainly investigated and found to result in improved MEA durability while maintaining similar MEA performance, compared to that observed with water–IPA or NMP only. In addition, a relationship between the performance-durability and average ionomer particle size was found. Larger and smaller particle sizes of ionomers in dispersions caused better performance and durability, respectively. Hence, the average particle size of ionomers in a dispersion is a good indicator to predict the performance and durability of MEAs in PEMFCs.

Author Contributions: Conceptualization, J.-S.P.; methodology, J.-S.P.; experimentation, C.-H.S.; validation, J.-S.P. and C.-H.S.; investigation, J.-S.P. and C.-H.S.; resources, C.-H.S.; writing—original draft preparation, J.-S.P.; writing—review and editing, J.-S.P.; supervision, J.-S.P.

Funding: This research was funded by a 2016 Research Grant from Sangmyung University.

Acknowledgments: This work was supported by a 2016 Research Grant from Sangmyung University.

Conflicts of Interest: The authors declare no conflicts of interest.

References

1. Park, J.-S.; Shin, M.-S.; Kim, C.-S. Proton exchange membranes for fuel cell operation at low relative humidity and intermediate temperature: An updated review. *Curr. Opin. Electrochem.* **2017**, *5*, 43–55. [CrossRef]
2. Park, J.-S.; Choi, Y.-W. High durable anion-conducting ionomer binder formed by on-site crosslinking. *Chem. Lett.* **2013**, *42*, 998–1000. [CrossRef]
3. Song, C.-H.; Park, J.-S. Membrane-electrode assemblies with patterned electrodes for proton exchange membrane fuel cells. *Chem. Lett.* **2018**, *47*, 196–199. [CrossRef]
4. Radev, I.; Koutzarov, K.; Pfrang, A.; Tsotridis, G. The influence of the membrane thickness on the performance and durability of PEFC during dynamic aging. *Int. J. Hydrogen Energy* **2012**, *37*, 11862–11870. [CrossRef]
5. Kim, K.-H.; Lee, K.-Y.; Kim, H.-J.; Cho, E.; Lee, S.-Y.; Lim, T.-H.; Yoon, S.P.; Hwang, I.C.; Jang, J.H. The effects of Nafion® ionomer content in PEMFC MEAs prepared by a catalyst-coated membrane (CCM) spraying method. *Int. J. Hydrogen Energy* **2010**, *35*, 2119–2126. [CrossRef]
6. Jeon, S.; Lee, J.; Rios, G.M.; Kim, H.-J.; Lee, S.-Y.; Cho, E.; Lim, T.-H.; Jang, J.H. Effect of ionomer content and relative humidity on polymer electrolyte membrane fuel cell (PEMFC) performance of membrane-electrode assemblies (MEAs) prepared by decal transfer method. *Int. J. Hydrogen Energy* **2010**, *35*, 9678–9686. [CrossRef]
7. Suzuki, A.; Sen, U.; Hattori, T.; Nagumo, R.; Tsuboi, H.; Hatakeyama, N.; Takaba, H.; Williams, M.C.; Miyamoto, A. Ionomer content in the catalyst layer of polymer electrolyte membrane fuel cell (PEMFC): Effects on diffusion and performance. *Int. J. Hydrogen Energy* **2011**, *36*, 2221–2229. [CrossRef]
8. Uchida, M.; Aoyama, Y.; Eda, N.; Ohta, A. Investigation of the microstructure in the catalyst layer and effects of both perfluorosulfonate ionomer and PTFE-loaded carbon on the catalyst layer of polymer electrolyte fuel cells. *J. Electrochem. Soc.* **1995**, *142*, 4143–4149. [CrossRef]
9. Paganin, V.A.; Ticianelli, E.A.; Gonzalez, E.R. Development and electrochemical studies of gas diffusion electrodes for polymer electrolyte fuel cells. *J. Appl. Electrochem.* **1996**, *26*, 297–304. [CrossRef]
10. Antolini, E.; Giorgi, L.; Pozio, A.; Passalacqua, E. Influence of Nafion loading in the catalyst layer of gas-diffusion electrodes for PEFC. *J. Power Sources* **1999**, *77*, 136–142. [CrossRef]
11. Park, Y.-C.; Tokiwa, H.; Kakinuma, K.; Watanabe, M.; Uchida, M. Effects of carbon supports on Pt distribution, ionomer coverage and cathode performance for polymer electrolyte fuel cells. *J. Power Sources* **2016**, *315*, 179–191. [CrossRef]
12. Lopez-Haro, M.; Guétaz, L.; Printemps, T.; Morin, A.; Escribano, S.; Jouneau, P.-H.; Bayle-Guillemaud, P.; Chandezon, F.; Gebel, G. Three-dimensional analysis of Nafion layers in fuel cell electrodes. *Nat. Commun.* **2014**, *5*, 5529. [CrossRef] [PubMed]
13. Lee, M.; Uchida, M.; Yano, H.; Tryk, D.A.; Uchida, H.; Watanabe, M. New evaluation method for the effectiveness of platinum/carbon electrocatalysts under operating conditions. *Electrochim. Acta* **2010**, *55*, 8504–8512. [CrossRef]

14. Kim, T.-H.; Yi, J.-Y.; Jung, C.-Y.; Jeong, E.; Yi, S.-C. Solvent effect on the Nafion agglomerate morphology in the catalyst layer of the proton exchange membrane fuel cells. *Int. J. Hydrogen Energy* **2017**, *42*, 478–485. [CrossRef]
15. Orfanidi, A.; Rheinländer, P.J.; Schulte, N.; Gasteiger, H.A. Ink solvent dependence of the ionomer distribution in the catalyst layer of a PEMFC. *J. Electrochem. Soc.* **2018**, *165*, F1254–F1263. [CrossRef]
16. Jung, C.-Y.; Yi, S.-C. Improved polarization of mesoporous electrodes of a proton exchange membrane fuel cell using N-methyl-2-pyrrolidinone. *Electrochim. Acta* **2013**, *113*, 37–41. [CrossRef]
17. Zhu, S.; Wang, S.; Jiang, L.; Xia, Z.; Sun, H.; Sun, G. High Pt utilization catalyst prepared by ion exchange method for direct methanol fuel cells. *Int. J. Hydrogen Energy* **2012**, *37*, 14543–14548. [CrossRef]
18. Omata, T.; Tanaka, M.; Miyatake, K.; Uchida, M.; Uchida, H.; Watanabe, M. Preparation and fuel cell performance of catalyst layers using sulfonated polyimide ionomers. *Appl. Mater. Interfaces* **2012**, *4*, 730–737. [CrossRef]
19. Rolfi, A.; Oldani, C.; Merlo, L.; Facchi, D.; Ruffo, R. New perfluorinated ionomer with improved oxygen permeability for application in cathode polymeric electrolyte membrane fuel cell. *J. Power Sources* **2018**, *396*, 95–101. [CrossRef]
20. Stassi, A.; Gatto, I.; Passalacqua, E.; Antonucci, V.; Aricò, A.S.; Merlo, L.; Oldani, C.; Pagano, E. Performance comparison of long and short-side chain perfluorosulfonic membranes for high temperature polymer electrolyte membrane fuel cell opersion. *J. Power Sources* **2011**, *196*, 8925–8930. [CrossRef]
21. Welch, C.; Labouriau, A.; Hjelm, R.; Orler, B.; Johnston, C.; Kim, Y.S. Nafion in dilute solvent systems: Dispersion or solution? *ACS Macro Lett.* **2012**, *1*, 1403–1407. [CrossRef]
22. Johnston, C.M.; Lee, K.-S.; Rockward, T.; Labouriau, A.; Mack, N.; Kim, Y.S. Impact of solvent on ionomer structure and fuel cell durability. *ECS Trans.* **2009**, *25*, 1617–1622.
23. Huang, D.-C.; Yu, P.-J.; Liu, F.-J.; Huang, S.-L.; Hsueh, K.-L.; Chen, Y.-C.; Wu, C.-H.; Chang, W.-C.; Tsau, F.-H. Effect of dispersion solvent in catalyst ink on proton exchange membrane fuel cell performance. *Int. J. Electrochem. Sci.* **2011**, *6*, 2551–2565.
24. Choi, B.; Langlois, D.A.; Mack, N.; Johnston, C.M.; Kim, Y.S. The effect of cathode structures on Nafion membrane durability. *J. Electrochem. Soc.* **2014**, *161*, F1154–F1162. [CrossRef]
25. Fuel Cell Tech Team Accelerated Stress Test and Polarization Curve Protocols for PEM Fuel Cells. Available online: https://www.energy.gov/sites/prod/files/2015/08/f25/fcto_dwg_usdrive_fctt_accelerated_stress_tests_jan2013.pdf (accessed on 28 November 2018).
26. Paul, D.K.; Karan, K.; Docoslis, A.; Giorgi, J.B.; Pearce, J. Characteristics of self-assembled ultrathin Nafion films. *Macromolecules* **2013**, *46*, 3461–3475. [CrossRef]
27. Karan, K. PEFC catalyst layer: Recent advances in materials, microstructural characterization, and modeling. *Curr. Opin. Electrochem.* **2017**, *5*, 27–35. [CrossRef]
28. Liu, M.; Wang, C.; Zhang, J.; Wang, J.; Hou, Z.; Mao, Z. Diagnosis of membrane electrode assembly degradation with drive cycle test technique. *Int. J. Hydrogen Energy* **2014**, *39*, 14370–14375. [CrossRef]
29. Kim, S.M.; Ahn, C.-Y.; Cho, Y.-H.; Kim, S.; Hwang, W.; Jang, S.; Shin, S.; Lee, G.; Sung, Y.-E.; Choi, M. High-performance fuel cell with stretched catalyst-coated membrane: One-step formation of cracked electrode. *Sci. Rep.* **2016**, *6*, 26503–26509. [CrossRef]

© 2019 by the authors. Licensee MDPI, Basel, Switzerland. This article is an open access article distributed under the terms and conditions of the Creative Commons Attribution (CC BY) license (http://creativecommons.org/licenses/by/4.0/).

Article

Experimental Studies of Effect of Land Width in PEM Fuel Cells with Serpentine Flow Field and Carbon Cloth

Xuyang Zhang *, Andrew Higier, Xu Zhang and Hongtan Liu

Clean Energy Research Institute, College of Engineering, University of Miami, Coral Gables, FL 33146, USA; ahigier@gmail.com (A.H.); xuzhang@miami.edu (X.Z.); hliu@miami.edu (H.L.)
* Correspondence: x.zhang18@umiami.edu; Tel.: +1-305-284-3916

Received: 26 December 2018; Accepted: 31 January 2019; Published: 1 February 2019

Abstract: Flow field plays an important role in the performance of proton exchange membrane (PEM) fuel cells, such as transporting reactants and removing water products. Therefore, the performance of a PEM fuel cell can be improved by optimizing the flow field dimensions and designs. In this work, single serpentine flow fields with four different land widths are used in PEM fuel cells to study the effects of the land width. The gas diffusion layers are made of carbon cloth. Since different land widths may be most suitable for different reactant flow rates, three different inlet flow rates are studied for all the flow fields with four different land widths. The effects of land width and inlet flow rate on fuel cell performance are studied based on the polarization curves and power densities. Without considering the pumping power, the cell performance always increases with the decrease in the land width and the increase in the inlet flow rates. However, when taking into consideration the pumping power, the net power density reaches the maximum at different combinations of land widths and reactant flow rates at different cell potentials.

Keywords: proton exchange membrane fuel cell; flow field; flow field design

---

1. Introduction

Proton exchange membrane (PEM) fuel cells are one of the promising renewable energy devices, owing to its high energy efficiency, low operating temperature, zero emission, and low noise. However, there are still some barriers that inhibit the widespread application of PEM fuel cells, such as high cost, low power density, and low durability [1,2]. Wilberforce et al. [3] reviewed the development of fuel cell electric cars and Zhang et al. [4,5] reviewed the degradation mechanisms in PEM fuel cell, and they found that the cost reduction, performance, and durability should be improved for the commercialization of the fuel cell. Flow field plays an important role in the performance and durability of PEM fuel cells since it helps to transport the reactants to the catalyst layer and reduces the liquid water accumulation in the catalyst layer [6,7]. Jung et al. [8] found that hydrogen crossover was strongly affected by the flow field dimension of the anode side. Oluwatosin et al. [9] reviewed different types of materials for flow field, and they found that a suitable coating on the flow field plate improved the corrosion resistance and fuel cell performance. There are three most commonly used flow fields: Serpentine, parallel, and interdigitated flow fields. The serpentine flow field is the most commonly used, since it has a better mass transfer than parallel flow field and a lower pumping power than interdigitated flow field [10].

Although the under-land cross-flow induced by the pressure difference between two adjacent channels in serpentine flow field helps to remove the excess produced liquid water and enhances the mass transfer, the concentration loss can still be high and can affect the fuel cell performance significantly. Many researchers have studied the channel designs from their numerical models [11].

Owejan et al. [12] found that the mass transfer loss can be decreased when the channel cross-section is changed from a rectangular shape to a triangular shape. Wang et al. [13] found that when the channel aspect ratio (channel height/width) decreased, the performance under both medium and low operating potential increased due to the increase of the under-land cross-flow rate, but the pumping power was increased as well. In addition, they found that the single serpentine flow field had a higher under-land cross-flow rate than multi-pass serpentine flow field. However, Nam et al. [14] found that the multi-pass serpentine flow-field had a higher under-land cross-flow rate than the single serpentine flow field instead, and the uniformity of local conditions and the fuel cell performance were improved with the multi-pass serpentine flow field. Shimpalee et al. [15] found that the 26-channel serpentine flow field gave the best performance and the lowest pumping power. Suresh et al. [16] introduced a split serpentine flow field with enhanced under-land cross-flow, and it improved the fuel cell performance significantly. Zheng et al. [17] added baffles in the downstream of the channel to improve the oxygen concentration and water removal ability, thus the performance under high current density region was increased.

Flow rate plays a significant role in fuel cell performance. Hu et al. [18] found that the flow rate had a large effect on the reactant concentration distribution in fuel cell. Wilberforce et al. [19] found that increasing the cathode flow rate increased the fuel cell performance, since it relieved the water flooding. Andrew et al. [20] found that the high flow rate increased the local current density under the land and channel areas in the fuel cell. However, a high flow rate means a high pressure drop [21], as well as a high pumping power that can lead to a low net power output, thus a balance between flow rate and pumping power is necessary.

The dimension of the land in flow fields also plays an important role in PEM fuel cell performance. A large land width ensures a good electrical and thermal conduction and high performance under high potentials, whereas a narrow land width provides high under-land cross-flow rate and high water removal capability [10,22]. Therefore, the land width should be optimized to reach a high performance for PEM fuel cells. Yoon et al. [23] studied the effect of land width on fuel cell performance under different relative humidity, but the type of flow fields used was not mentioned. Akhtar et al. [24] found that a smaller land-to-channel width ratio led to a more uniform profile of liquid water saturation and oxygen consumption in interdigitated flow fields. Cooper et al. [25] found that the land-to-channel width ratio played a more important role in performance for both parallel and interdigitated flow fields from their experimental results. Liu et al. [26] conducted experimental tests to optimize the total channel-land width and the land-to-channel width ratio in serpentine flow fields, and they found that the relatively small total widths of lands and channels, together with a small land-to-channel width ratio, provides the highest performance.

In the above literature reviews, we can see that the flow field dimension affects the PEM fuel performance significantly, but the pumping power and flow rate are rarely considered at the same time. Without considering the pumping power, the improvement of performance caused by optimizing the flow field dimension can be incorrect in real-life PEM fuel cells. Carbon cloth is used as the gas diffusion layer (GDL) in fuel cells and has a high permeability and a high water removal capability than carbon paper [1,27], but the experimental studies with carbon cloths are very limited. In this work, the effect of land width of serpentine flow field on fuel cell performance is experimentally studied with four different land widths and three different inlet flow rates, and the carbon cloth is used as the GDL. In addition, the net power densities under different potentials that include the pumping power are also evaluated for different flow rates and different land widths.

2. Experimental Section

2.1. Fuel Cell Test System

Hydrogenics® test system G60 with build-in automatic software HywareII™ is used to control the operating potential, measure the current density (by RBL electrical load, ±0.5% accuracy), and then plot the polarization curves. The schematic diagram of the experimental test system is shown in Figure 1. The test station controls the anode/cathode inlet temperature and relative humidity and flow rates (by Bronkhost mass flow controller, ±0.5% accuracy), back pressure and operating temperature. When the reactant gas flows through the bubble humidifiers and preheater in the test system, the inlet gas temperature and dew point are controlled, thus the relative humidity of inlet gas is controlled as well. The inlet pressure and outlet pressure on both anode and cathode can be measured with separate pressure transducers (Omega, ±0.08% accuracy) in the test system.

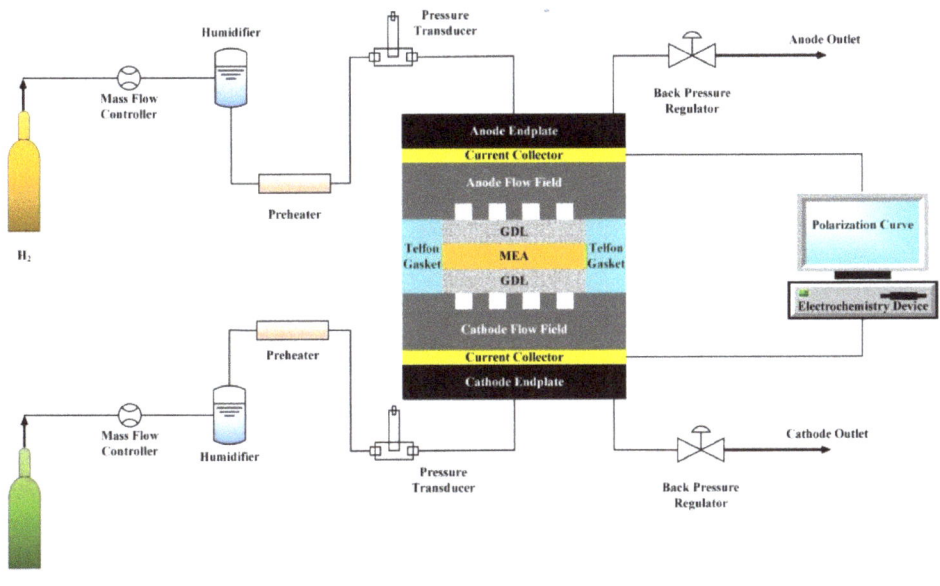

Figure 1. Schematic diagram of Hydrogenics® test system G60.

2.2. Fixture and Operation Conditions

The membrane used in the fuel cell is Nafion™ 117 purchased from Alfa-Aesar. The carbon cloth electrodes for the anode and cathode sides are purchased from BCS Fuel Cells. The platinum catalyst loading is 0.4 mg cm^2. The membrane electrode assemblies (MEA) is assembled and hot pressed in-house. The MEA with an electrode area of 44 cm^2 is assembled by two end-plates, two gold-coated copper plates, two graphite plates with 44 cm^2 single serpentine flow fields, and two Teflon® gaskets. The gold-coated copper plates are used to collect the current from the cell. The Teflon® gaskets are used to avoid gas leakage and control the compression ratio of GDL thickness. The single serpentine flow field used at the anode side has 1 mm channel depth, 1 mm channel width, and 1 mm land width. The four single serpentine flow fields used at the cathode side all have 1 mm channel depth, 1 mm channel, but their land widths are 0.5 mm, 1 mm, 1.5 mm, and 2 mm, respectively, shown as Figure 2.

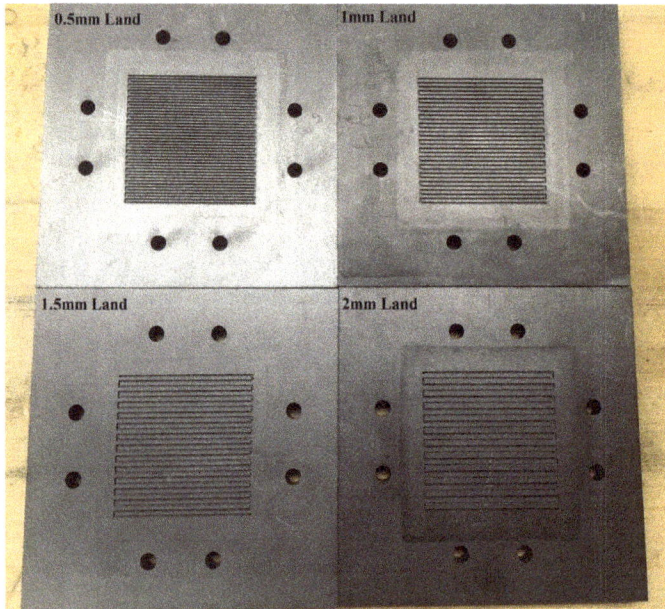

Figure 2. Single serpentine flow fields with four different land widths (0.5 mm, 1 mm, 1.5 mm, 2 mm).

A series of experiments with three different air inlet flow rates of 0.5, 1, and 2 L/min are conducted at the cathode side using four different cathode flow fields, respectively. The hydrogen inlet flow rate at the anode side is kept at 0.5 L/min. The cell operating temperature, inlet gas temperatures of the anode and cathode sides of the fuel cells are maintained at 70 °C. The dew point of both anode and cathodes gases are kept at 70 °C to ensure fully humidified gases. No back pressure is applied in the experiments, and the outlet pressure on both anode and cathode is kept at ambient pressure.

Before the actual experiment, a break-in procedure with MEA humidification and catalyst activation is applied. In the present work, in obtaining the polarization curves, the voltage is changed from open-circuit potential (OCV) of the cell to 0.25 V with a step of 0.05 V. At each voltage stage, the current and voltage values are collected only after the cell has reached stead-state conditions.

3. Results and Discussions

3.1. Influence of Inlet Flow Rates

Figure 3 shows the current density of different operating potentials at different inlet flow rates. As the flow rate increases, the current densities at less than 0.6 V cell potentials, always increase in all the four different land width cases. This result is consistent with the findings in the previous literature [20]. The reason is that the available reactant gas to the catalyst layer increases and excess produced liquid water can be removed easily at a higher inlet flow rate. Therefore, the mass transfer loss is lower at a higher inlet flow rate.

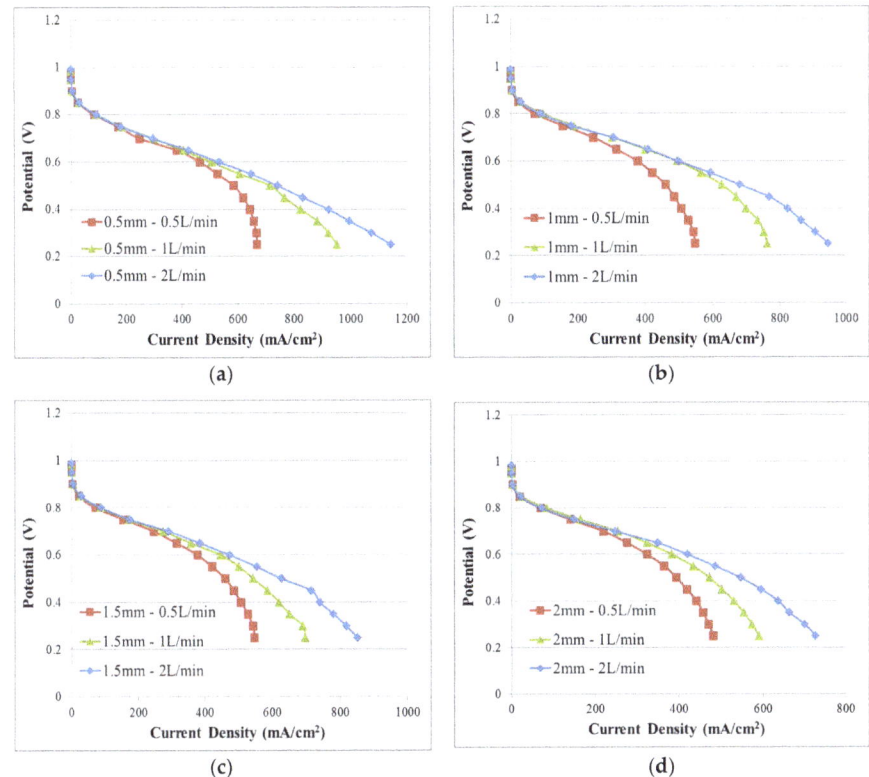

Figure 3. Comparison of polarization curves at different inlet flow rates: (**a**) Land width = 0.5 mm; (**b**) land width = 1 mm; (**c**) land width = 1.5 mm; (**d**) land width = 2 mm.

The pressure drop from the inlet to the outlet at high flow rates is higher, and this can lead to a higher pumping power. The pumping power (W_p) can be calculated from the pressure drop and inlet flow rate from the following equation.

$$W_p = \frac{PQ_{in}}{\eta} \tag{1}$$

where W_p is the pumping power, P is the pressure drop from the inlet to the outlet, Q_{in} is the inlet flow rate, and η is the pumping efficiency and assumed to be 0.8 in this study. The average pressure drop from the inlet to the outlet at different inlet flow rates during the fuel cell operating are listed in Table 1. The pressure drop increases significantly with the inlet flow rate due to the high gas velocity along the channel and more produced liquid water. The pressure drop decreases significantly when the land width increases from 0.5 mm to 1 mm, since the total channel length is decreased significantly for a fixed flow field area. However, when the land width further increases from 1 mm, the pressure drop does not vary too much with the decrease of the total channel length, since the low under-land cross-flow rate for larger land widths cannot effectively remove the excess product liquid water, and accumulated liquid water in the flow field increases the pressure drop. Since the inlet flow rate on the anode is kept at a low level and no liquid water is produced at the anode side, the pressure drop at the anode side is low and the pumping power at the anode side is not considered in this study.

Table 1. Average pressure drop at different inlet flow rates for four different serpentine flow fields at the cathode side.

Pressure drop (kPa)	0.5 L/min	1 L/min	2 L/min
0.5 mm Land	66.35	132.30	259.28
1 mm Land	42.21	85.43	186.72
1.5 mm Land	40.58	83.57	176.98
2mm Land	41.57	83.69	176.98

Figure 4 displays the net power densities at different operating potentials and different inlet flow rates after considering the pumping power. It can be seen that the net power density does not vary too much at 0.7 V for all the four flow fields when the inlet flow rate increases from 0.5 L/min to 1 L/min. At a high cell potential, the high flow rate increases the amount of reactant gas available to the catalyst layer, but this benefit is just enough to compensate the high pumping power. The maximum net power density occurs at 0.6 V and 0.5 V when the inlet flow rate is at 1 L/min, owing to a suitable balance between the amount of reactant gas to the catalyst layer and the pumping power. At 0.4 V, the net power density does not vary too much from 1 L/min to 2 L/min, except for the case with 0.5 mm land width. At a low cell potential, current density is high and a large amount of liquid water is produced, and the mass transfer loss becomes an important factor for fuel cell performance. The high inlet flow rate decreases the mass transfer loss, and this can compensate the high pumping power at a low cell potential, i.e., 0.4 V. Therefore, when the pumping power is considered, there is an optimal inlet flow rate to maximize the fuel cell performance for each operating condition, listed in Table 2. Normally, the optimal inlet flow rate is 1 L/min for these four serpentine flow fields, instead of the maximum flow rate as the literature reported [20].

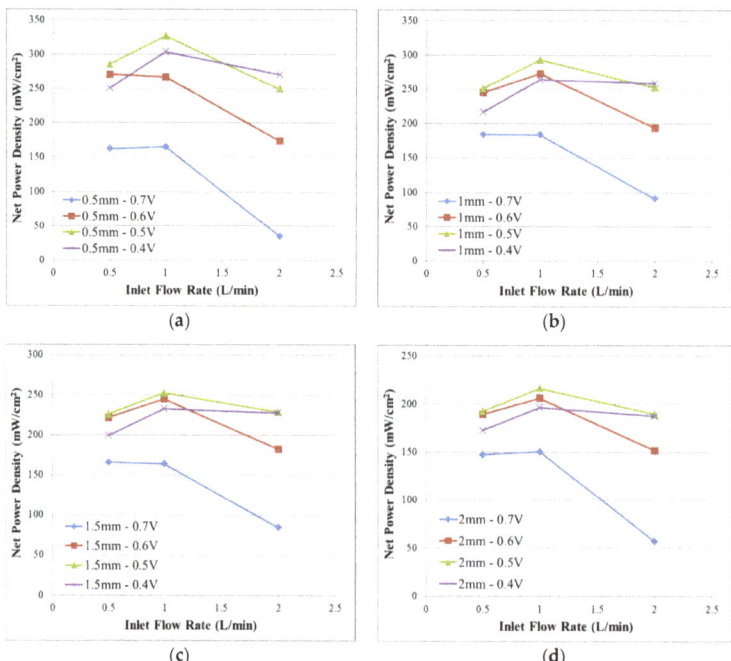

Figure 4. Comparison of net power densities that include the pumping power at different inlet flow rates and different land widths: (**a**) Land width = 0.5 mm; (**b**) land width = 1 mm; (**c**) land width = 1.5 mm; (**d**) land width = 2 mm.

Table 2. Optimal inlet flow rate for four single serpentine flow fields with different land widths at different cell potentials when the pumping power is considered.

Optimized Inlet Flow Rate (L/min)	0.5 mm	1 mm	1.5 mm	2 mm
0.7 V cell potential	0.5 or 1	0.5 or 1	0.5 or 1	0.5 or 1
0.6 V cell potential	0.5 or 1	1	1	1
0.5 V cell potential	1	1	1	1
0.4 V cell potential	1	1 or 2	1 or 2	1 or 2

3.2. Influence of Land Width

Figure 5 shows the performance of PEM fuel cell with different land widths. As the land width decreases from 2 mm to 0.5 mm, the current densities always increase when the cell potentials are at 0.6 V or lower for all the inlet flow rates. Even at a low inlet flow rate, i.e., 0.5 L/min, the high under-land cross-flow rate caused by decreasing the land width still significantly increases the water removal capability and significantly decreases the mass transfer loss, and this phenomenon is consistent with the findings in the literature [28]. The single serpentine flow field with 0.5 mm land width always has the best performance when the pumping power is not considered. These results agreed with the previous literature [23,26], as the decrease in the land width leads to an increase of fuel cell performance.

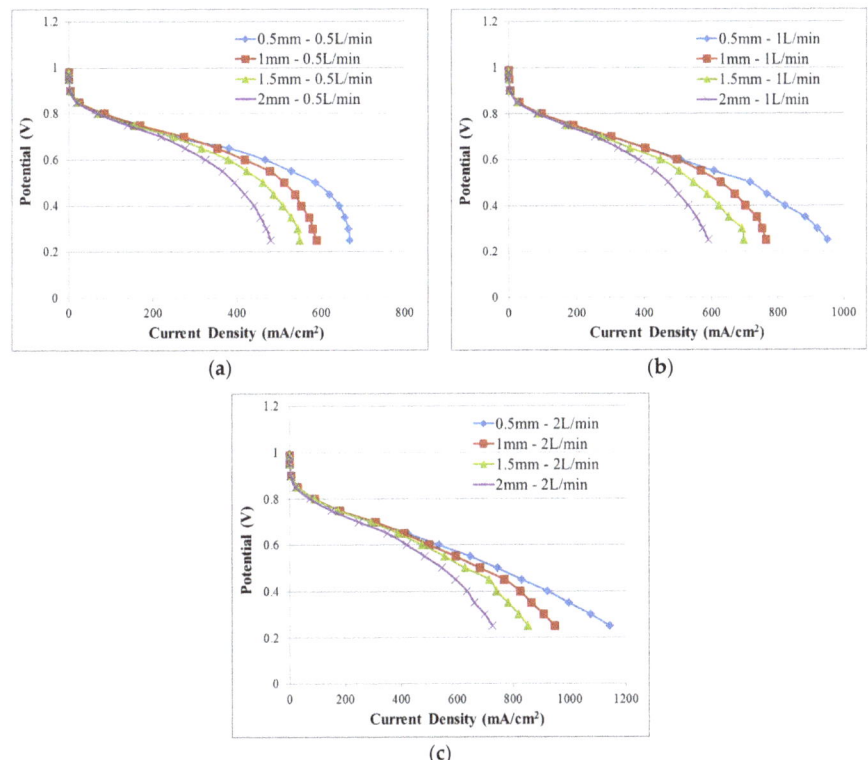

Figure 5. Comparison of polarization curves at different land widths: (**a**) Inlet flow rate = 0.5 L/min; (**b**) inlet flow rate = 1 L/min; (**c**) inlet flow rate = 2 L/min.

Figure 6 shows the net power densities with different land widths taking into consideration of the pumping power. At a low inlet flow rate (e.g., 0.5 L/min), the net power densities for the 0.5 mm land width case out-perform the other cases when the cell potential is 0.6 V or lower. At a medium inlet flow rate (e.g., 1 L/min), the net power densities for the 0.5 mm land width case out-perform the other cases when the cell potential is 0.5 V or lower. At a high inlet flow rate (i.e., 2 L/min), the net power densities for the 0.5 mm land width case out-perform the other cases only when the cell potential is at 0.4 V. Thus, when the inlet flow rate is higher, the higher under-land cross-flow rate caused by a narrower land only benefits the fuel cell performance at the lower cell potential region. The reason is that the high inlet flow rate leads to a higher pumping power. As a result, the benefit of narrow land width on fuel cell performance is lower at higher inlet flow rates for single serpentine flow fields when the pumping power is considered. Table 3 lists the optimal land width at different inlet flow rates and different cell potentials, and the narrowest land width (i.e., 0.5 mm) does not always result in the best performance when the pumping power is considered. As a result, the optimal flow field by decreasing the land width as the literature stated [23,26] is not valid when pumping power is considered.

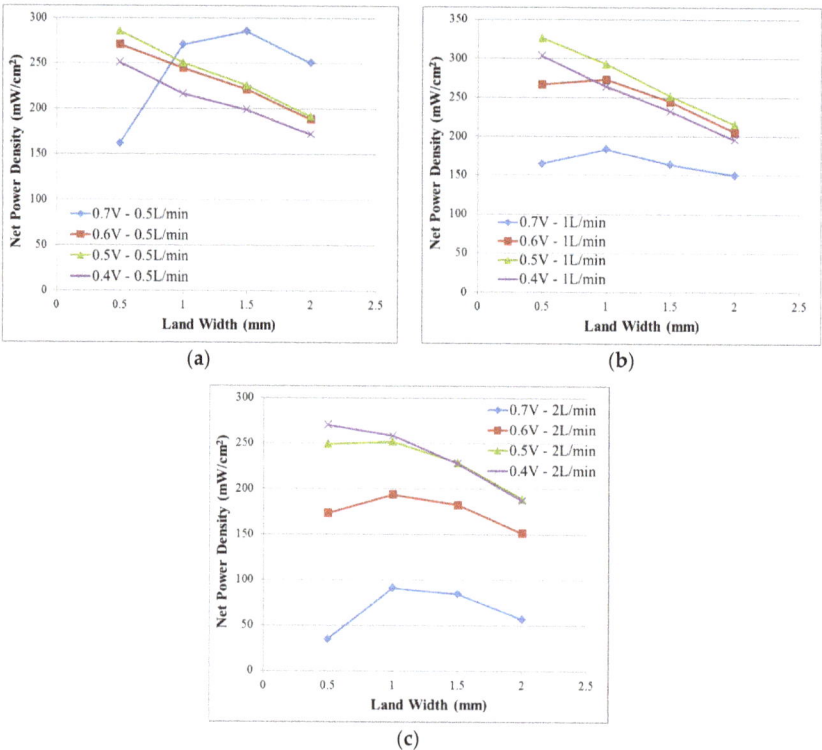

Figure 6. Comparison of net power densities that include the pumping power at different land widths: (**a**) Inlet flow rate = 0.5 L/min; (**b**) inlet flow rate = 1 L/min; (**c**) inlet flow rate = 2 L/min.

Table 3. Optimal land width in single serpentine flow fields at different inlet flow rates and different cell potentials when the pumping power is considered.

Optimized Land Width (mm)	0.5 L/min	1 L/min	2 L/min
0.7 V cell potential	1.5	1	1
0.6 V cell potential	0.5	1	1
0.5 V cell potential	0.5	0.5	0.5 or 1
0.4 V cell potential	0.5	0.5	0.5

4. Conclusions

Four single serpentine flow fields with different land widths and three inlet flow rates are used in the experiment to study the effect of land width on fuel cell performance. Fuel cell performance for different land widths was evaluated both with and without the consideration of the pumping power. The following conclusions can be made based on the experimental results:

(1) For all the cases with different land width, high inlet flow rate (i.e., 2 L/min) always resulted in high fuel cell performance when the pumping power was not considered, whereas the medium inlet flow rate (i.e., 1 L/min) generally provided the highest fuel cell performance when the pumping power was included.

(2) As the land width decreases, the fuel cell performance without considering the pumping power at 0.6 V cell potential or lower always increases, due to the increase of the under-land cross-flow rate and decrease of the mass transfer loss.

(3) When the pumping power was included, the improvement of fuel cell performance caused by the decrease of the land width only occurred at a lower cell potential at a higher inlet flow rate.

(4) The effects of the land width and inlet flow rate on fuel cell performance when considering the pumping power are very different from that without considering the pumping power.

(5) Without considering the pumping power, the improvement of performance caused by optimizing the land width can be over-estimated, and the inlet flow rate cannot be optimized properly.

Author Contributions: Conceptualization, X.Z. (Xuyang Zhang) and H.L.; methodology, A.H.; experiment and data collection, A.H. and X.Z. (Xuyang Zhang); data analysis, X.Z. (Xuyang Zhang); investigation, X.Z. (Xuyang Zhang) and X.Z. (Xu Zhang); resources, H.L.; writing—original draft preparation, X.Z. (Xuyang Zhang); writing—review and editing, X.Z. (Xuyang Zhang), X.Z. (Xu Zhang) and H.L.; project administration, H.L.

Funding: This research received no external funding.

Conflicts of Interest: The authors declare no conflict of interest.

References

1. O'hayre, R.; Cha, S.-W.; Prinz, F.B.; Colella, W. *Fuel Cell Fundamentals*; John Wiley & Sons: Hoboken, NJ, USA, 2016.
2. Laramie, J.; Dicks, A. *Fuel Cell Systems Explained*; John Wiley and Sons: New York, NY, USA, 2003.
3. Wilberforce, T.; El-Hassan, Z.; Khatib, F.N.; Al Makky, A.; Baroutaji, A.; Carton, J.G.; Olabi, A.G. Developments of electric cars and fuel cell hydrogen electric cars. *Int. J. Hydrog. Energy* **2017**, *42*, 25695–25734. [CrossRef]
4. Zhang, S.; Yuan, X.-Z.; Hin, J.N.C.; Wang, H.; Friedrich, K.A.; Schulze, M. A review of platinum-based catalyst layer degradation in proton exchange membrane fuel cells. *J. Power Sources* **2009**, *194*, 588–600. [CrossRef]
5. Zhang, S.; Yuan, X.; Wang, H.; Merida, W.; Zhu, H.; Shen, J.; Wu, S.; Zhang, J. A review of accelerated stress tests of MEA durability in PEM fuel cells. *Int. J. Hydrog. Energy* **2009**, *34*, 388–404. [CrossRef]
6. Li, X.; Sabir, I. Review of bipolar plates in PEM fuel cells: Flow-field designs. *Int. J. Hydrog. Energy* **2005**, *30*, 359–371. [CrossRef]
7. Tawfik, H.; Hung, Y.; Mahajan, D. Metal bipolar plates for PEM fuel cell—A review. *J. Power Sources* **2007**, *163*, 755–767. [CrossRef]
8. Jung, A.; Kong, I.M.; Baik, K.D.; Kim, M.S. Crossover effects of the land/channel width ratio of bipolar plates in polymer electrolyte membrane fuel cells. *Int. J. Hydrog. Energy* **2014**, *39*, 21588–21594. [CrossRef]
9. Ijaodola, O.; Ogungbemi, E.; Khatib, F.N.; Wilberforce, T.; Ramadan, M.; Hassan, Z.E.; Thompson, J.; Olabi, A.G. Evaluating the Effect of Metal Bipolar Plate Coating on the Performance of Proton Exchange Membrane Fuel Cells. *Energies* **2018**, *11*, 3203. [CrossRef]
10. Kahraman, H.; Orhan, M.F. Flow field bipolar plates in a proton exchange membrane fuel cell: Analysis & modeling. *Energy Convers. Manag.* **2017**, *133*, 363–384.
11. Vinh, N.; Kim, H.-M. Comparison of Numerical and Experimental Studies for Flow-Field Optimization Based on Under-Rib Convection in Polymer Electrolyte Membrane Fuel Cells. *Energies* **2016**, *9*, 844. [CrossRef]

12. Owejan, J.; Trabold, T.; Jacobson, D.; Arif, M.; Kandlikar, S. Effects of flow field and diffusion layer properties on water accumulation in a PEM fuel cell. *Int. J. Hydrog. Energy* **2007**, *32*, 4489–4502. [CrossRef]
13. Wang, X.-D.; Duan, Y.-Y.; Yan, W.-M.; Lee, D.-J.; Su, A.; Chi, P.-H. Channel aspect ratio effect for serpentine proton exchange membrane fuel cell: Role of sub-rib convection. *J. Power Sources* **2009**, *193*, 684–690. [CrossRef]
14. Nam, J.H.; Lee, K.-J.; Sohn, S.; Kim, C.-J. Multi-pass serpentine flow-fields to enhance under-rib convection in polymer electrolyte membrane fuel cells: Design and geometrical characterization. *J. Power Sources* **2009**, *188*, 14–23. [CrossRef]
15. Shimpalee, S.; Greenway, S.; Van Zee, J.W. The impact of channel path length on PEMFC flow-field design. *J. Power Sources* **2006**, *160*, 398–406. [CrossRef]
16. Suresh, P.V.; Jayanti, S.; Deshpande, A.P.; Haridoss, P. An improved serpentine flow field with enhanced cross-flow for fuel cell applications. *Int. J. Hydrog. Energy* **2011**, *36*, 6067–6072. [CrossRef]
17. Zhang, G.; Fan, L.; Sun, J.; Jiao, K. A 3D model of PEMFC considering detailed multiphase flow and anisotropic transport properties. *Int. J. Heat Mass Transf.* **2017**, *115*, 714–724. [CrossRef]
18. Hu, X.; Wang, X.; Chen, J.; Yang, Q.; Jin, D.; Qiu, X. Numerical Investigations of the Combined Effects of Flow Rate and Methanol Concentration on DMFC Performance. *Energies* **2017**, *10*, 1094. [CrossRef]
19. Wilberforce, T.; El-Hassan, Z.; Khatib, F.N.; Al Makky, A.; Baroutaji, A.; Carton, J.G.; Thompson, J.; Olabi, A.G. Modelling and simulation of Proton Exchange Membrane fuel cell with serpentine bipolar plate using MATLAB. *Int. J. Hydrog. Energy* **2017**, *42*, 25639–25662. [CrossRef]
20. Higier, A.; Liu, H. Direct measurement of current density under the land and channel in a PEM fuel cell with serpentine flow fields. *J. Power Sources* **2009**, *193*, 639–648. [CrossRef]
21. Wilberforce, T.; El-Hassan, Z.; Khatib, F.N.; Al Makky, A.; Mooney, J.; Barouaji, A.; Carton, J.G.; Olabi, A.-G. Development of Bi-polar plate design of PEM fuel cell using CFD techniques. *Int. J. Hydrog. Energy* **2017**, *42*, 25663–25685. [CrossRef]
22. Manso, A.P.; Marzo, F.F.; Barranco, J.; Garikano, X.; Garmendia Mujika, M. Influence of geometric parameters of the flow fields on the performance of a PEM fuel cell. A review. *Int. J. Hydrog. Energy* **2012**, *37*, 15256–15287. [CrossRef]
23. Yoon, Y.-G.; Lee, W.-Y.; Park, G.-G.; Yang, T.-H.; Kim, C.-S. Effects of channel configurations of flow field plates on the performance of a PEMFC. *Electrochim. Acta* **2004**, *50*, 709–712. [CrossRef]
24. Akhtar, N.; Kerkhof, P.J.A.M. Effect of channel and rib width on transport phenomena within the cathode of a proton exchange membrane fuel cell. *Int. J. Hydrog. Energy* **2011**, *36*, 5536–5549. [CrossRef]
25. Cooper, N.J.; Smith, T.; Santamaria, A.D.; Park, J.W. Experimental optimization of parallel and interdigitated PEMFC flow-field channel geometry. *Int. J. Hydrog. Energy* **2016**, *41*, 1213–1223. [CrossRef]
26. Liu, H.; Li, P.; Wang, K. Optimization of PEM fuel cell flow channel dimensions—Mathematic modeling analysis and experimental verification. *Int. J. Hydrog. Energy* **2013**, *38*, 9835–9846. [CrossRef]
27. Wang, Y.; Wang, C.-Y.; Chen, K.S. Elucidating differences between carbon paper and carbon cloth in polymer electrolyte fuel cells. *Electrochim. Acta* **2007**, *52*, 3965–3975. [CrossRef]
28. Jiao, K.; Park, J.; Li, X. Experimental investigations on liquid water removal from the gas diffusion layer by reactant flow in a PEM fuel cell. *Appl. Energy* **2010**, *87*, 2770–2777. [CrossRef]

© 2019 by the authors. Licensee MDPI, Basel, Switzerland. This article is an open access article distributed under the terms and conditions of the Creative Commons Attribution (CC BY) license (http://creativecommons.org/licenses/by/4.0/).

Article

Liquid Water Transport in Porous Metal Foam Flow-Field Fuel Cells: A Two-Phase Numerical Modelling and Ex-Situ Experimental Study

Ashley Fly [1], Kyoungyoun Kim [2], John Gordon [1], Daniel Butcher [1] and Rui Chen [1,*]

[1] Department of Aeronautical and Automotive Engineering, Loughborough University, Loughborough LE11 3TU, UK; A.Fly@lboro.ac.uk (A.F.); J.D.A.Gordon2@lboro.ac.uk (J.G.); D.Butcher@lboro.ac.uk (D.B.)

[2] Department of Mechanical Engineering, Hanbat National University, Daejeon 34158, South Korea; KKim@hanbat.ac.kr

* Correspondence: R.Chen@lboro.ac.uk

Received: 27 February 2019; Accepted: 26 March 2019; Published: 27 March 2019

Abstract: Proton exchange membrane fuel cells (PEMFCs) using porous metallic foam flow-field plates have been demonstrated as an alternative to conventional rib and channel designs, showing high performance at high currents. However, the transport of liquid product water through metal foam flow-field plates in PEMFC conditions is not well understood, especially at the individual pore level. In this work, ex-situ experiments are conducted to visualise liquid water movement within a metal foam flow-field plate, considering hydrophobicity, foam pore size and air flow rate. A two-phase numerical model is then developed to further investigate the fundamental water transport behaviour in porous metal foam flow-field plates. Both the experimental and numerical work demonstrate that unlike conventional PEMFC channels, air flow rate does not have a strong influence on water removal due to the high surface tensions between the water and foam pore ligaments. A hydrophobic foam was seen to transport liquid water away from the initial injection point faster than a hydrophilic foam. In ex-situ tests, liquid water forms and maintains a random preferential pathway until the flow-field edge is reached. These results suggest that controlled foam hydrophobicity and pore size is the best way of managing water distribution in PEMFCs with porous flow-field plates.

Keywords: PEMFC; metal foam; channel; flow-field; water transport; mass transport; two-phase; numerical model

1. Introduction

The flow-field plate in a proton exchange membrane fuel cell (PEMFC), often referred to as the flow-field, must effectively serve multiple functions simultaneously to achieve good cell performance. The ideal flow-field plate should have a high electrical and thermal conductivity for electron and heat transport, have good mechanical and chemical stability, provide even reactant distribution across the cell active area and maintain membrane humidity whilst facilitating the removal of product water [1]. The most commonly used flow-field plate for PEMFCs is the rib and channel design, consisting of flow channels pressed, machined or etched into a conductive plate, as shown in Figure 1a. The channel area allows for reactant gas distribution and product water removal, whereas the rib area is in contact with the gas diffusion layer (GDL) to provide mechanical strength and electron transport to the external circuit. The design of traditional flow channels has received significant attention in the literature, with many numerical and experimental studies focusing on flow channel layout [2], shape [3] and wall hydrophobicity [4].

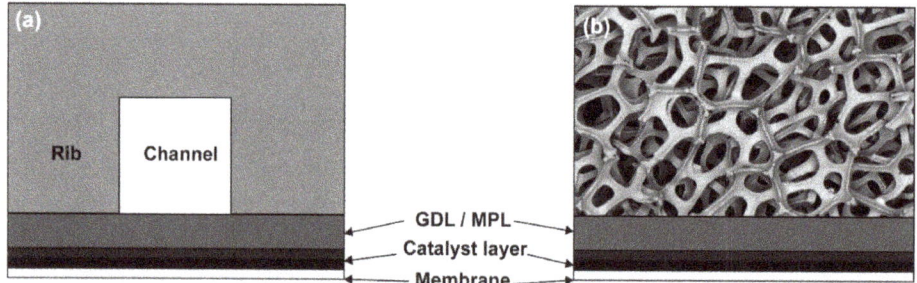

Figure 1. Cross section illustration of (**a**) conventional rib and channel flow-field and (**b**) porous metal foam flow-field.

Numerical modelling shows that to minimise excess water saturation, a hydrophobic GDL should be combined with a hydrophilic straight flow channel to transport water from the GDL to the flow channel where it can be more easily removed [5,6]. However, at the bends in serpentine flow channels Song et al. [4] demonstrated that a hydrophobic channel wall is best for minimising liquid saturation of the GDL.

A disadvantage of the rib and channel design is that the area under the rib experiences a higher flow resistance than the channel, leading to localised reduced reactant concentrations and product water build up. This behaviour was observed experimentally through neutron beam imaging by Meyer et al. [7]. Interdigitated flow channels, where flow is forced through the GDL underneath the ribs, can be used to improve water removal, but at the expense of increased pressure drop caused by the lower permeability of the GDL [3]. Several alternative flow-field plates for PEMFCs have been proposed in the literature, including: bio inspired [8], sintered metal [9], 3D metal mesh [10] and porous metal foams [11–16]. Of these, porous metal foam flow-field plates (Figure 1b) have seen the greatest interest due to both low cost and high productions volumes due to extensive use as battery electrodes [17].

Murphy et al. [11] was the first to use porous metal foam as the flow-field plate in a PEMFC, testing an eight cell stack and conducting single phase flow distribution tests using dyed liquid water. Kumar and Reddy [12] then conducted a direct comparison between multi-parallel channels and a Ni–Cr foam flow-field plate, showing improved mass transport behaviour in the foam flow-field at high currents. Similar behaviour was also observed by Tseng et al. [13], Tsai et al. [14], Shin et al. [15] and Kim et al. [18], with the improvement performance believed to be due to improved reactant distribution of the metal foam flow-field, as demonstrated during single phase flow visualisation by Fly et al. [19]. Fly et al. [20] investigated how foam compression affects fuel cell performance through electrochemical tests and X-ray computed tomography, demonstrating improved performance up to 70% thickness compression, primarily due to improved contact between the foam and bi-polar plate.

Whilst there have been several studies on the electrochemical performance of metal foam flow-fields in fuel cells, the two-phase gas and water interaction within the foam flow-field is not as well understood. Tabe et al. [16] conducted the first in-situ visualisation of liquid water behaviour in a PEMFC with foam flow-fields using a transparent endplate. The study found that hydrophilic coated foam provided both higher and more stable voltage than a hydrophobic coated foam because of the foam's ability to draw liquid away from the GDL.

Outside of the fuel cell literature, two-phase flow in metal foam has been studied with application to packed columns in the chemical processing industry [21,22]. Both Calvo et al. [21] and Wallenstein et al. [22] performed X-ray tomography on metal/ceramic foam columns subjected to gas and liquid in counter flow. Both authors observed static hold up; the volume fraction of liquid remaining after gas and liquid flows were stopped in the region of 5–10%. However, both the foam geometry and flow conditions differed from those seen in fuel cell flow-fields.

In the present work, liquid water transport in metal foams subject to two-phase PEMFC flow conditions is investigated through a combination of ex-situ visualisation experiments and computational fluid dynamics numerical modelling. Different foam geometry, hydrophobicity and flow rates are considered and methods for minimising liquid water saturation on the GDL surface are investigated.

2. Experimental Methodology

An ex-situ representation of a section of metal foam flow-field was designed to visualise liquid water transport through the foam in a controlled environment independent of the electrochemical reaction. The test section consisted of a 40 mm × 10 mm sample of metal foam surrounded by a silicone gasket and sandwiched between a set of 6.0 mm thick clear acrylic plates. A 1.0 mm diameter hole was laser cut into the 1.0 mm thick acrylic sheet on the bottom surface of the foam, through which liquid water was delivered to represent product water entering the flow-field from the GDL. Liquid water was delivered to the hole through a channel cut into a 1.5 mm thick acrylic plate. Dry air flow was supplied in plane with the foam sheet, perpendicular to the liquid flow. The two-phase mixing point was located at least 10 foam pore lengths from the foam inlet to eliminate entrance effects; Figure 2 illustrates the test apparatus used.

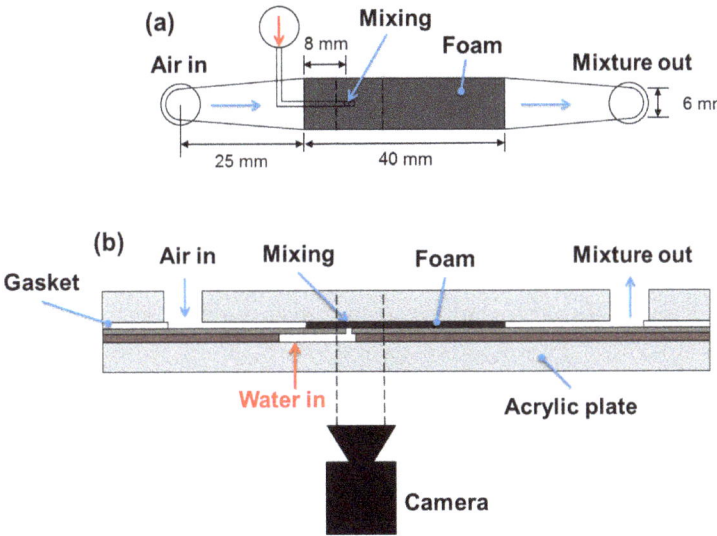

Figure 2. Ex-situ test fixture design (**a**) Top view, (**b**) Side section view through mixing point.

Liquid water was supplied to the sample at a flow rate of 3.0 gh^{-1}, controlled using a high-resolution needle valve and liquid mass flow meter (Bronkhorst, Netherlands). To better optically separate the water from the air, blue dye was added to the water at a water to dye ratio of 286:1. The low concentration of dye was not deemed to influence the properties of the water. Air flow rates of 0.5 and 2.0 Lmin^{-1} were used, regulated using a mass flow controller (Bronkhorst, Netherlands). This corresponded to air velocities ranging from 0.42 to 3.32 ms^{-1} at the foam inlet, depending on the foam thickness used. A hydrophobic coating (Electrolube Fluorocoat) was applied to the acrylic plate with the hole to better represent the surface of the GDL and numerical model conditions.

Two different foams were tested: a 1.6 mm thick nickel foam (Foam A, Corun New Energy, China) and a 3.0 mm thick nickel foam (Foam B, Sumitomo, Japan). Table 1 shows the manufacturer properties of the foams tested where pore size is the effective diameter of an average pore. The foams were paired with 1.0 mm and 3.0 mm thick Shore 60 A silicone gaskets, respectively, and the fixture was compressed using 12 bolts tightened to 3 Nm. Water transport through the foam was captured using a Cannon EOS 100D SLR camera with frame rate of 60 frames per second and resolution 1280 × 720. The camera was attached to an Infinity K2 distamax long-distance optic microscope. All tests were performed at room temperature (22–25 °C).

Table 1. Metal foam properties.

Parameter	Foam A	Foam B
Thickness	1.6 mm	3.0 mm
Material	Nickel	Nickel
Pore size	0.23 mm	0.95 mm
Porosity	97%	95%

3. Numerical Model

The flow in the metal foam was assumed to be unsteady, isothermal and laminar three-dimensional flow. In this study, the volume of fluid (VOF) model was employed for the two-phase flow simulation in the gas channel. The governing equations for the two-phase flow are the continuity equation and the Navier-Stokes equation:

$$\frac{\partial \rho}{\partial t} + \nabla \cdot (\rho \mathbf{V}) = 0 \tag{1}$$

$$\frac{\partial \rho \mathbf{V}}{\partial t} + \nabla \cdot (\rho \mathbf{V}\mathbf{V}) = -\nabla p + \nabla \cdot \left(\mu\left(\nabla \mathbf{V} + \nabla \mathbf{V}^T\right)\right) + \rho \mathbf{g} + \mathbf{F} \tag{2}$$

where p is pressure, ρ and μ are volume-averaged density and viscosity, respectively, $\mathbf{V}$ is fluid velocity and $\mathbf{g}$ is gravitational acceleration. $\mathbf{F}$ represents the momentum source term associated with the surface tension and is expressed according to the continuum surface force (CSF) model [23] as follows:

$$\mathbf{F} = \sigma \kappa_i \frac{\rho \nabla \alpha_i}{<\rho>} \tag{3}$$

where σ is the surface tension coefficient, $\langle \rho \rangle$ is the average density of the two fluids and the curvature at the interface κ_i is calculated from the local gradient of the surface normal vector $\mathbf{n} = \nabla \alpha_i / |\nabla \alpha_i|$:

$$\kappa_i = \nabla \cdot \mathbf{n}. \tag{4}$$

The volume fraction of fluid i, α_i, was calculated in every computational cell over the entire domain:

$$\frac{\partial \alpha_i}{\partial t} + \mathbf{V} \cdot \nabla \alpha_i = 0. \tag{5}$$

The wall adhesion was taken into account by imposing the unit normal vector of the interface at the wall as:

$$\mathbf{n} = \mathbf{n}_w \cos\theta + \mathbf{t}_w \sin\theta \tag{6}$$

where $\mathbf{n}_w$ and $\mathbf{t}_w$ are unit vectors normal and tangential to the solid wall, respectively and θ is the contact angle.

Figure 3a shows the computational domain used in this study. The domain size was 1.9 mm × 2.8 mm × 0.9 mm in the streamwise, spanwise length and height, respectively. The tetrakaidecahedron (or Kelvin cell) was used to model the pore geometry in the metal foam, which had 14 faces (6 quadrilateral and 8 hexagonal) and 24 vertices. The pore size was 0.345 mm and the ligament diameter was set as 0.054 mm to make the porosity of 0.95. These values are representative of commercial foam geometry utilised in the literature for metal foam flow-fields [15,18]. To model emerging droplets on the GDL surface, liquid water was injected through a hydrophobic bottom wall. The size of the pores through which the liquid droplets emerged was 50 μm × 50 μm. The bottom surface of the channel and the surface of the metal foam were assumed to be hydrophobic and the contact angle of 110° was imposed on those surfaces.

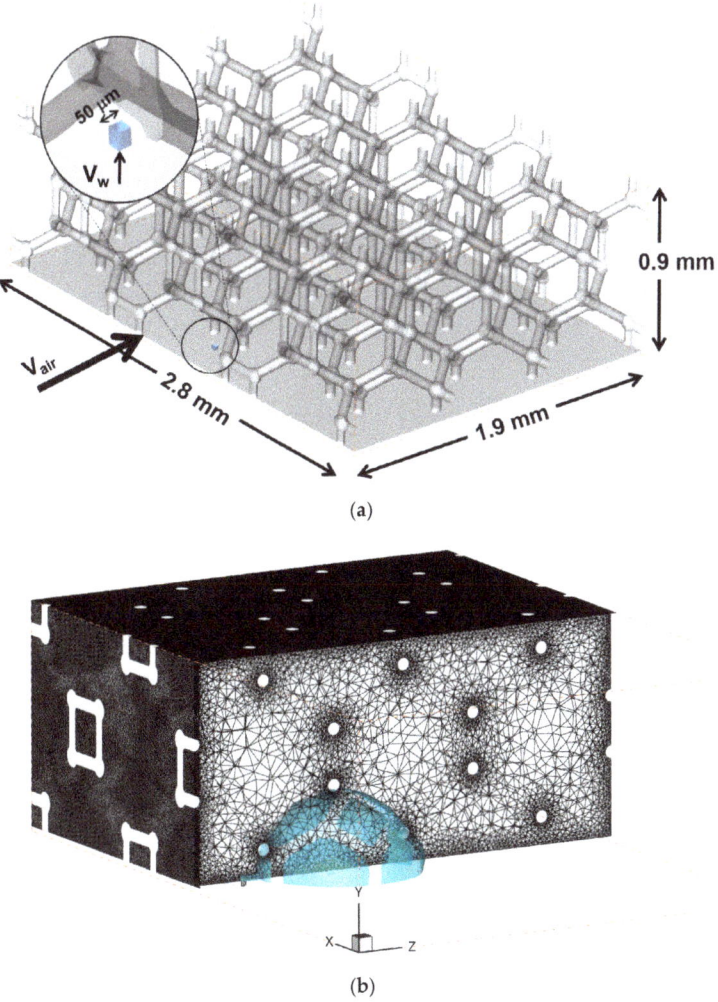

Figure 3. Numerical model (**a**) Computational domain, (**b**) Mesh.

In the air inlet and water inlet, a constant velocity was applied, and the outflow condition was used for the outlet. The boundary conditions except the inlet and outlet were set to no-slip condition. V_{air} denotes the airflow velocity into the flow channel at a constant speed of 2.0, 3.0 and 4.0 ms^{-1} and V_w is the liquid water velocity supplied through the pores at a constant speed of 1.66 ms^{-1} to represent the product water exiting the GDL of a PEMFC. The governing equations were solved by the interface tracking between the pressure–velocity coupling method through the PISO scheme and the gas-liquid through the geo-reconstruct scheme. In the geometric reconstruction scheme, the interface between air and water was determined by a piecewise linear interface calculation method. The time-advancement was made with the time step of 1.0×10^{-6} s by the first-order implicit scheme. The residual of each governing equation was 0.001. The simulations have been carried out using the commercial CFD code

ANSYS Fluent software in conjunction with the built-in CSF model [24]. The number of grid cells used in the calculation was about 5,500,000. Figure 3b shows the computational mesh at a plane through the water inlet. Although the simulations were performed on 128-core Linux clusters, the wall clock CPU time was required about one month for time marching of 1 s, which limited the time extent of the present analysis to the very early phase.

4. Analysis and Discussion

4.1. Experimental Results

Ex-situ visualisation of water transport in the metal foam flow-field was conducted using the test setup described in Section 2. Figure 4 shows how the liquid water moved through foam A during the first 120 s from initial liquid injection at the mixing point for both the 0.5 Lmin^{-1} and 2.0 Lmin^{-1} air flow rate. Air flow was from left to right, gravity was acting in the direction out of the page, and liquid flow rate was 3.0 gh^{-1}. Increasing the air flow rate increased the rate at which the water front moved through the foam flow-field in the streamwise direction, reducing the time taken to remove product water from the flow-field in a working fuel cell.

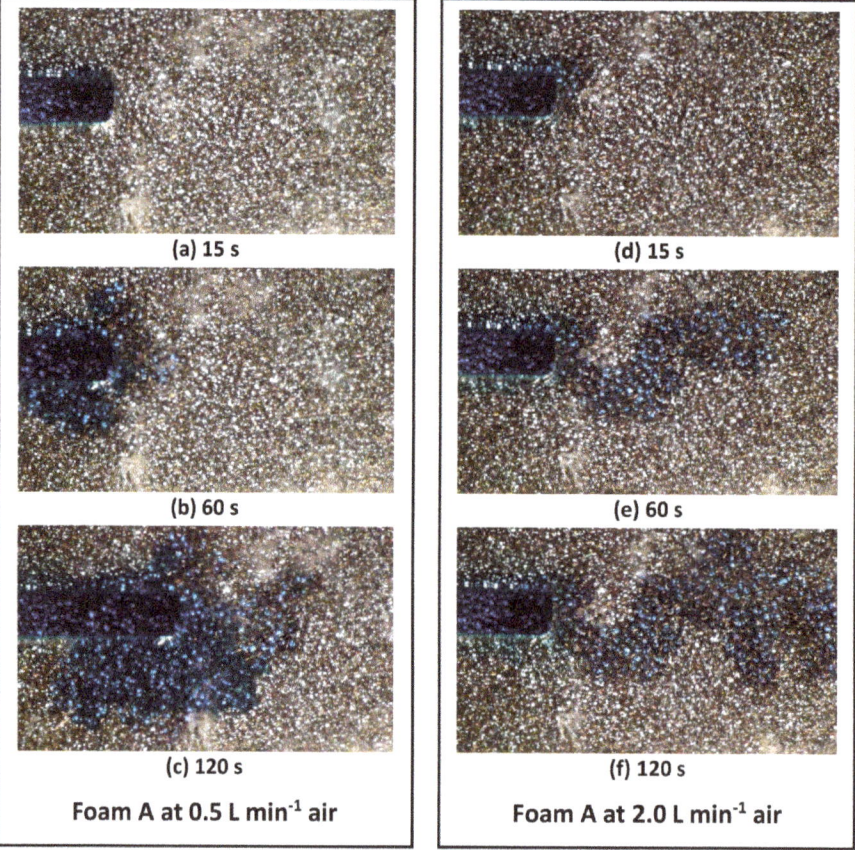

Figure 4. Propagation of liquid water in foam A (1.6 mm thick) air flow (left to right), 7.7 mm × 5.0 mm field of view. After (**a**) 15 s, (**b**) 60 s and (**c**) 120 s at 0.5 L min^{-1} air flow and (**d**) 15 s, (**e**) 60 s and (**f**) 120 s at 2.0 L min^{-1} air flow.

For both flow rates, the liquid water front was seen to travel upstream of the airflow (left of initial mixing point) a short distance before spreading out and moving downstream in the flow direction. This was more pronounced in the 0.5 Lmin^{-1} case and can be clearly seen in Figure 4b. The fluid behaviour relating to the upstream movement after initial liquid injection was investigated further using the numerical modelling results in Section 4.2.

After initial upstream movement, the water front spread more evenly away from the mixing point until the width of saturated foam (top to bottom of image) was sufficient enough that the force of the airflow overcame the water surface tension in the foam pores and the water front moved in the airflow direction. At higher air flow rates, the width of the water front required before upstream movement dominated was reduced as surface tension forces were more readily overcome by the increased air momentum, as seen by comparing Figure 4c,f.

Throughout both tests, the water was seen to establish and maintain a preferential pathway in the airflow direction, with all additional water then travelling along the same pathway to reach the edge of the foam sample. This contrasted with the transport mechanism in a conventional fuel cell flow channel where droplets are formed and then detached from the pore, travelling along the flow channel in a periodic manner [4]. An established water pathway within the fuel cell flow-field was beneficial in removing product water quickly; however, a wide water channel, such as that seen in Figure 4c, could prevent reactant gases reaching the active area, causing localised fuel starvation.

Each test condition in Figure 4 was repeated three times and the foam removed and dried between tests. In each case the same behaviour was observed but a different water pathway was established due to the non-homogenous construction of the foam. The initial movement of water upstream of the airflow also occurred when the air flow direction was reversed (right to left), negating effects of the fixture design and orientation, and was not seen to occur when the tests were repeated with the foam sample removed. The flow penetration length was calculated and presented in Figure 5.

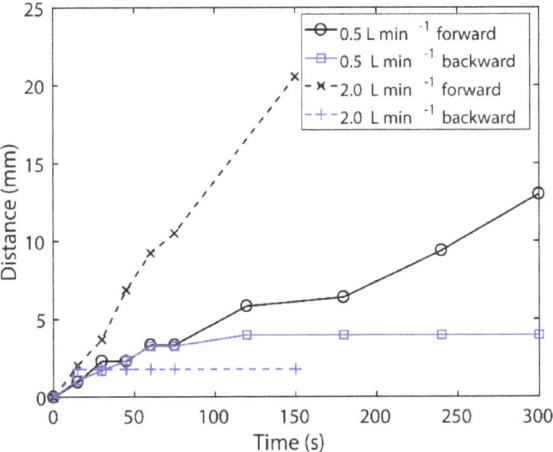

Figure 5. Liquid water penetration length for foam A.

The water propagation in foam B was presented in Figure 6 for the same conditions and field of view as Figure 4. Water transport was seen to occur much faster in foam A, which had a 0.23 mm pore size, compared to foam B, which had a 0.95 mm pore size. The difference between water transport in the two foams was partially due to the increased thickness of foam B increasing the filling volume and reducing air velocity, but also due to the foam pore size. A smaller pore size, such as in foam A, increased the capillary effect of the porous media, making the movement of water from one foam pore to the next more likely to occur.

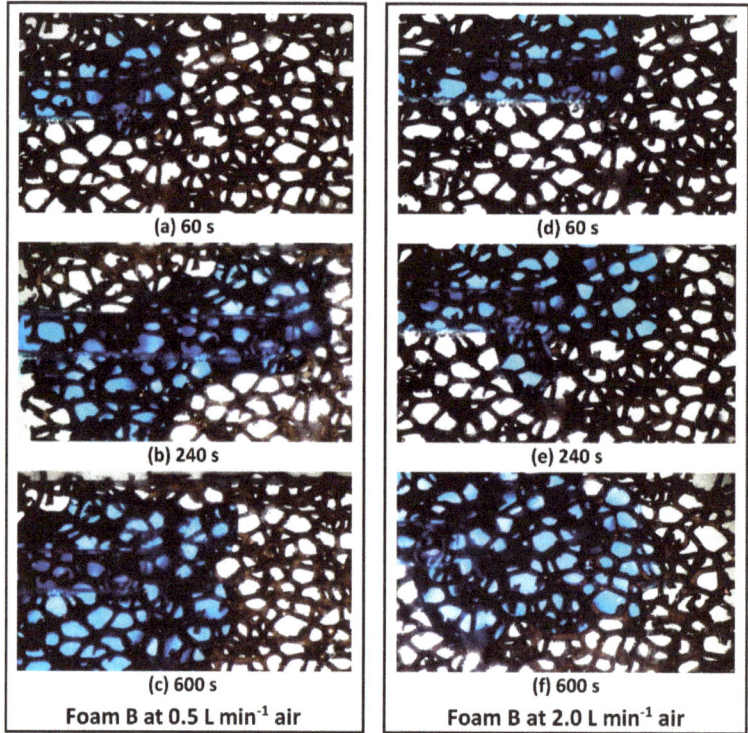

Figure 6. Propagation of liquid water in foam B (3.0 mm thick), air flow (left to left), 7.7 mm × 5.0 mm field of view. After (**a**) 60 s, (**b**) 240 s and (**c**) 600 s at 0.5 L min^{-1} air flow and (**d**) 60 s, (**e**) 240 s and (**f**) 600 s at 2.0 L min^{-1} air flow

In both foams it was observed that the water front only filled one foam pore at a time and not multiple pores simultaneously. Once a single foam pore was filled then the next pore would begin to fill. One potential mechanism is that the location of the next foam pore to be filled with water is based on the probability of the water overcoming the surface tension and air momentum effects in each individual pore. The non-homogeneous nature of the foam means that each foam pore had a different probability of filling. Increasing air flow rate increased the probability of the downstream foam pores filling as air momentum effects increased, as seen in Figures 4 and 6.

The distance of the water front both upstream and downstream of the mixing point over time is shown in Figures 5 and 7 for foams A and B, respectively. In foam B, backwards (upstream) propagation of the water front was seen to dominate for the first 200 s from initial mixing before forwards (downstream) movement occurred. In both foams the backwards water front remained stationary once forwards movement began to dominate. This means that unlike a conventional flow channel where water was effectively removed by the airflow, in a foam flow-field the water remained within the foam pore. In a functioning PEMFC, this behaviour could have the benefit of improved internal humidification and better proton conductivity, although excess water accumulation would cause localised fuel starvation events and increased air flow pressure drop. Furthermore, liquid trapped in foam pores would expand in sub-zero conditions, potentially damaging the fuel cell and effecting cold start behaviour.

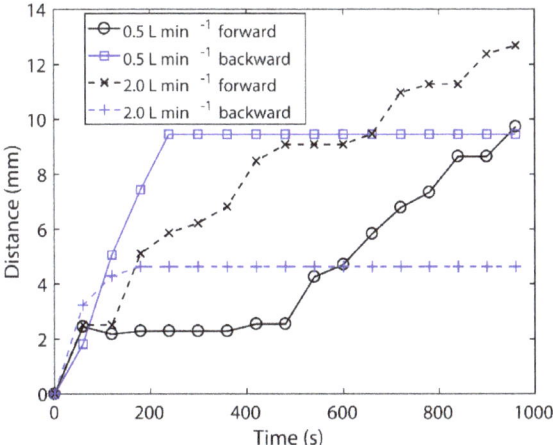

Figure 7. Liquid water penetration length for foam B.

To avoid stagnation of liquid water, the surface of foam B was given a hydrophobic coating by dipping the foam in a PTFE solution (25 wt %) then dried on a wire mesh for 2.5 hours at 100 °C, resulting in 18.8 wt % PTFE. The ex-situ experiments were then repeated with the PTFE coating; this hydrophobic treatment was not possible with foam A due to PTFE blocking the smaller pores. With the hydrophobic coating present on foam B, initial water transport moved away from the mixing point to the side edges of the flow channel, from where it travelled along the transition between the foam and gasket in the flow direction to the end of the sample. At 0.5 Lmin^{-1} the water reached the edge and end of the sample after 30 s and 550 s, respectively. For the 2.0 Lmin^{-1} air flow rate test, the water took 60 s to reach the edge and 260 s to reach the end of the foam sample. This compared to 1480 s and 1890 s to reach the end of the foam sample in the non-PTFE treated foam B sample at 0.5 Lmin^{-1} and 2.0 Lmin^{-1}, respectively. Time to liquid water breakthrough (reaching the end of the sample) for all test cases is shown in Table 2. This demonstrates that the addition of a hydrophobic coating to the foam significantly reduced the residence time for water transport, and hence the volume of water stored in the foam. Across a larger foam flow-field, as seen in a PEMFC, movement of water towards the flow-field edge caused by hydrophobic treated foams could be utilised as an effective water management strategy, for example, by having periodic separators in the foam, or utilising a wicking material around the edge of the flow-field. However, improper management could lead to areas of high water saturation on the extremities of the active area, leading to uneven current density distribution. Whilst every effort was made to ensure an even distribution of PTFE coating on foam B, inhomogeneous coating could occur due to fluid wicking, pore blocking or gravitational effects. Additional work is required in this area to ensure even PTFE coating across the surface of the metal foam ligaments.

Table 2. Time for liquid water to reach the end of the foam sample after initially entering at 3.0 gh^{-1}.

Foam	Air Flow Rate (Lmin^{-1})	Time for Liquid to Reach End of Sample (s)
Foam A	0.5	630
Foam A	2.0	195
Foam B	0.5	1480
Foam B	2.0	1890
Foam B (18.8 wt % PTFE)	0.5	550
Foam B (18.8 wt % PTFE)	2.0	260

4.2. Numerical Study Results

The numerical model described in Section 3 was used to facilitate better understanding of the fundamental flow behaviour occurring when liquid water entered the foam flow-field from the GDL, and to investigate the upstream flow behaviour seen in the experimental work.

The movement of the initial droplet after liquid injection with $V_{air} = 3$ ms^{-1} is shown in Figure 8. The liquid water droplets are visualised using the isosurface of $\alpha_w = 0.5$. Initially, the liquid water moved in the direction of the airflow (a), before suddenly decreasing in velocity (b), moving to the opposite direction of flow (c), and then adhering to adjacent metal foam (d), before the droplet size increased (e). The initial backward (upstream) movement of the droplet was consistently observed in simulations of different incoming flow velocity, as shown in Figure 9, confirming the same flow behaviour observed during the ex-situ experiments.

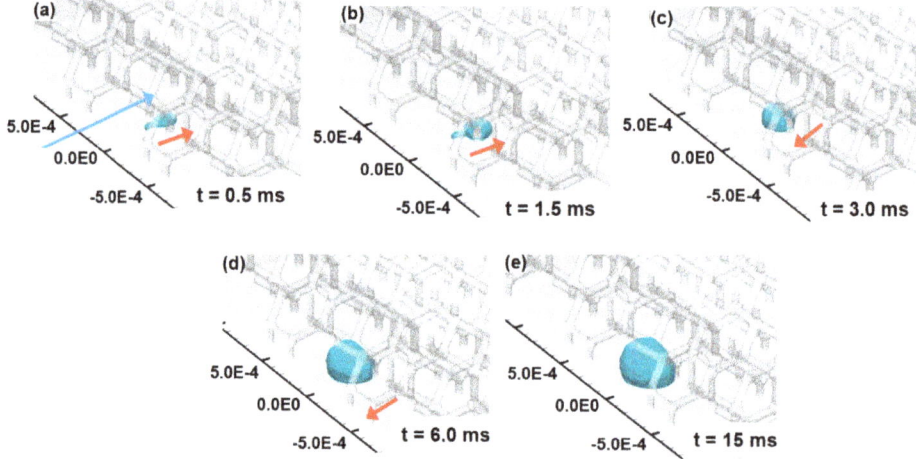

Figure 8. Initial behaviour of water droplet at 3 ms^{-1} air flow rate; (**a**) 0.5 ms, (**b**) 1.5 ms, (**c**) 3.0 ms, (**d**) 6.0 ms and (**e**) 15 ms after liquid water injection.

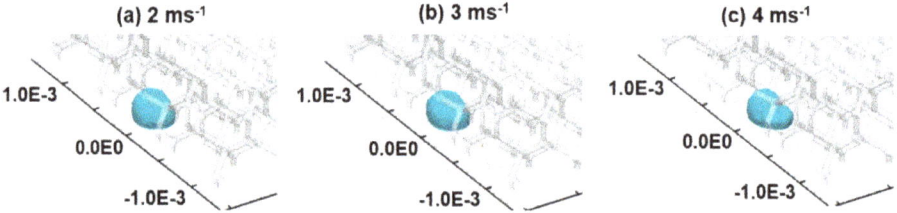

Figure 9. Comparison of water droplet propagation 20 ms after liquid injection for (**a**) 2 ms^{-1}, (**b**) 3 ms^{-1} and (**c**) 4 ms^{-1} air flow condition.

Figure 10 shows the velocity vector distribution around the droplet inside the metal foam. Surface tension caused droplets to attach to the ligaments and trap them inside the pores. It can be seen that most of the aerodynamic forces resulting from the inlet flow acted on the ligaments of the metal foam, so that the momentum of the inlet air flow could not be transferred to the droplet effectively and the droplet did not move well. This suggests that the surface tension was dominant compared with aerodynamic forces and thus the droplet dynamics in the metal foam having small pore size was mainly influenced by surface tension rather than the incoming air flow, as shown by the difference between Foam A and B in the experimental work and through Figure 9 in the numerical work. The backward propagation of liquid droplets observed in both the experimental and numerical work can be explained by greater adhesion force of adjacent upstream ligaments than aerodynamic drag exerted on the droplet. As the amount of water injected continued to increase over time, the droplet size increased.

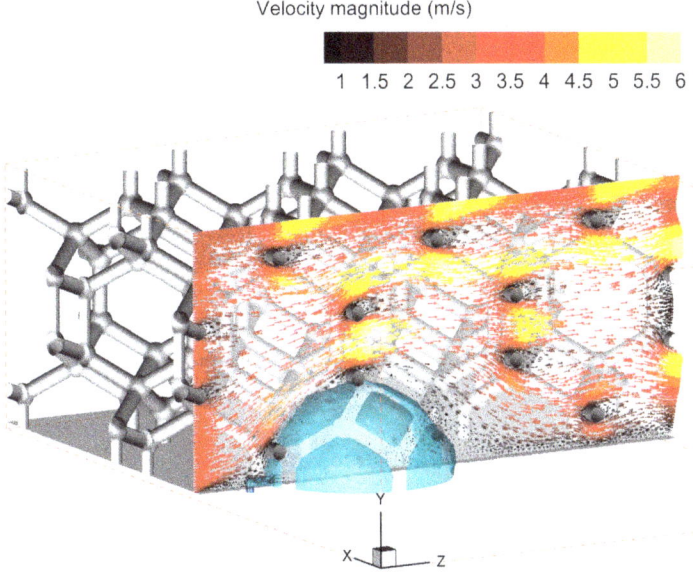

Figure 10. Distribution of velocity around the attached water droplet at t = 60 ms, air flow rate 3 ms^{-1} in hydrophobic foam.

The increased cross-sectional area made the aerodynamic drag acting on the droplet larger than the contact force, which led the droplet to be detached from the pore and to move in the flow direction. A new droplet was then attached to the different location of metal foam and coalesced with the second droplet that had been detached after it had grown up in the pore, forming a large droplet (Figure 11a). In a PEMFC environment this meant that once liquid water established a presence in the flow-field it was likely to be maintained until the droplet reached a critical volume where aerodynamic forces overcame surface tension adhesion force. Accumulation of liquid water on the GDL surface blocked reactant gas pathways and could lead to localised reactant gas starvation. However, despite this behaviour, many authors have observed improved performance from metal foam flow-fields compared to conventional rib and channel flow-fields during high current density operation. This is likely due to the area of the GDL being blocked by the flow-field rib in conventional designs (around 50%) being significantly greater than the area blocked by liquid water droplets in the metal foam.

The effect of modifying the metal foam hydrophobicity on water transport over multiple pores in the period several minutes after initial liquid addition was studied using the ex-situ experiments. To investigate the effect of the metal foam hydrophobicity on the droplet behaviour at the point of initial injection, VOF simulation was also performed for hydrophilic cases, in which the contact angle of the ligament was given as 70°. The initial droplet behaviour in the hydrophilic case Figure 11b was almost identical to the hydrophobic case Figure 11a. The hydrophilic case showed, however, that even if the size of the droplet increased as the water continued to be injected, the droplet was not separated, due to the hydrophilic ligament, and remained attached to the ligament. For both hydrophobic and hydrophilic cases, compared with the conventional flow channel, the liquid water droplet was not discharged efficiently, which was attributed to significant contact force of the metal foam due to the surface tension.

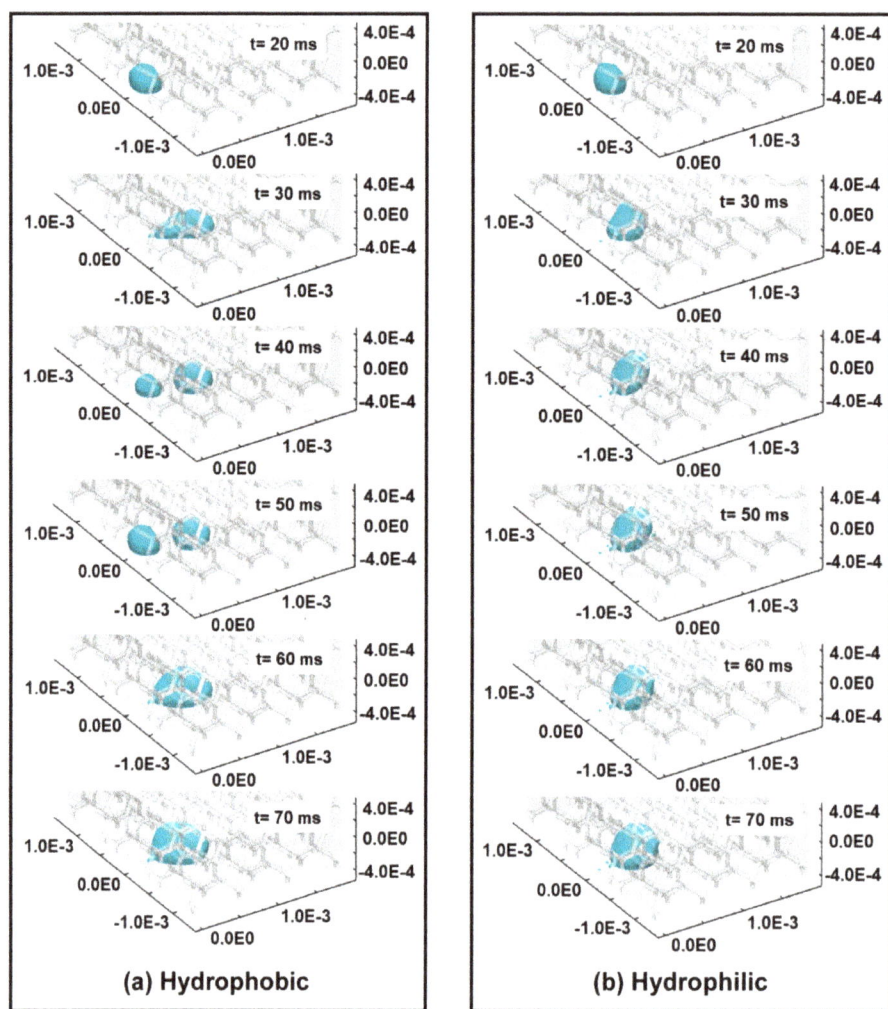

Figure 11. Effects of hydrophobicity on the droplet movement for airflow rate 3 ms^{-1} (**a**) hydrophobic and (**b**) hydrophilic.

5. Conclusions

Liquid water transport in a porous metal foam flow-field has been investigated through ex-situ visualisation and two-phase numerical modelling with application to PEMFCs. Both experimental and numerical results show that the transport of liquid water in a porous metal foam flow-field is more complex than a conventional rib and channel flow-field. Whilst investigating different length and time scales, the experimental and numerical results both demonstrate that adhesion forces though surface tension are larger than aerodynamic drag forces under PEMFC conditions. This causes upstream propagation of liquid water after initial injection and liquid water to accumulate in the foam pores until droplet size is sufficient to occupy an adjacent pore, or for aerodynamic drag to overcome surface tension.

A smaller foam pore size was seen to reduce the time taken for liquid water to propagate along the flow-field. Likewise, increasing the hydrophobicity of the foam was also seen to reduce the transport time of liquid water by promoting transport pathways along the edge of the flow-field. Both pore size and hydrophobicity had a greater influence on water transport than air flow rate.

The information obtained from this study can be used to inform future designs, materials and coating of porous flow-fields for improved water management and PEMFC performance.

Supplementary Materials: The underlying research used in this publication can be found at 10.17028/rd.lboro.7775843.

Author Contributions: Conceptualisation, R.C., K.K. and A.F.; methodology, K.K., A.F., J.G. and D.B.; software, K.K.; validation, A.F., J.G. and D.B.; writing—original draft preparation, A.F. and K.K.; writing—review and editing, A.F., K.K, J.G, D.B. and R.C; funding acquisition, R.C. and K.K.

Funding: This research was funded by the Engineering and Physical Sciences Research Council (EPSRC) under grant number EP/M023508/1 and EP/L015749/1, supported by the International Collaborative Energy Technology R&D Program of the Korea Institute of Energy Technology Evaluation and Planning (KETEP) and granted financial resources from the Ministry of Trade, Industry and Energy, Republic of Korea (No. 20148520120160).

Conflicts of Interest: The authors declare no conflict of interest. The funders had no role in the design of the study; in the collection, analyses, or interpretation of data; in the writing of the manuscript; or in the decision to publish the results.

References

1. Larminie, J.; Dicks, A. *Fuel Cell Systems Explained*; Wiley Online Library: Hoboken, NJ, USA, 2003; ISBN 047084857X.
2. Sasmito, A.P.; Kurnia, J.C.; Mujumdar, A.S. Numerical evaluation of various gas and coolant channel designs for high performance liquid-cooled proton exchange membrane fuel cell stacks. *Energy* **2012**, *44*, 278–291. [CrossRef]
3. Wang, C.; Zhang, Q.; Shen, S.; Xiaohui, Y.; Zhu, F.; Cheng, X. The respective effect of under-rib convection and pressure drop of flow fields on the performance of PEM fuel cells. *Nature* **2017**, *7*, 1–9. [CrossRef]
4. Song, M.; Kim, H.Y.; Kim, K. Effects of hydrophilic/hydrophobic properties of gas flow channels on liquid water transport in a serpentine polymer electrolyte membrane fuel cell. *Int. J. Hydrogen Energy* **2014**, *39*, 19714–19721. [CrossRef]
5. Kim, H.Y.; Jeon, S.; Song, M.; Kim, K. Numerical simulations of water droplet dynamics in hydrogen fuel cell gas channel. *J. Power Sources* **2014**, *246*, 679–695. [CrossRef]
6. Alrahmani, M.; Chen, R.; Ibrahim, S.; Patel, S. A Numerical Study on the Effects of Gas Channel Wettability in PEM Fuel Cells. *ECS Trans.* **2014**, *48*, 81–92. [CrossRef]
7. Meyer, Q.; Ashton, S.; Jervis, R.; Finegan, D.P.; Boillat, P.; Cochet, M.; Curnick, O.; Reisch, T.; Adcock, P.; Shearing, P.R.; et al. The Hydro-electro-thermal Performance of Air-cooled, Open-cathode Polymer Electrolyte Fuel Cells: Combined Localised Current Density, Temperature and Water Mapping. *Electrochim. Acta* **2015**, *180*, 307–315. [CrossRef]
8. Trogadas, P.; Cho, J.I.S.; Neville, T.P.; Marquis, J.; Wu, B.; Brett, D.J.L.; Coppens, M.-O. A lung-inspired approach to scalable and robust fuel cell design. *Energy Environ. Sci.* **2018**, *11*, 136–143. [CrossRef]
9. Kariya, T.; Hirono, T.; Funakubo, H.; Shudo, T. Effects of the porous structures in the porous flow field type separators on fuel cell performances. *Int. J. Hydrogen Energy* **2014**, *39*, 15072–15080. [CrossRef]
10. Yoshida, T.; Kojima, K. Toyota MIRAI Fuel Cell Vehicle and Progress Toward a Future Hydrogen Society. *Interface Mag.* **2015**, *24*, 45–49. [CrossRef]
11. Murphy, O.J.; Cisar, A.; Clarke, E. Low-cost light weight high power density PEM fuel cell stack. *Electrochim. Acta* **1998**, *43*, 3829–3840. [CrossRef]
12. Kumar, A.; Reddy, R.G. Materials and design development for bipolar/end plates in fuel cells. *J. Power Sources* **2004**, *129*, 62–67. [CrossRef]
13. Tseng, C.-J.; Tsai, B.T.; Liu, Z.-S.; Cheng, T.-C.; Chang, W.-C.; Lo, S.-K. A PEM fuel cell with metal foam as flow distributor. *Energy Convers. Manag.* **2012**, *62*, 14–21. [CrossRef]
14. Tsai, B.-T.; Tseng, C.-J.; Liu, Z.-S.; Wang, C.-H.; Lee, C.-I.; Yang, C.-C.; Lo, S.-K. Effects of flow field design on the performance of a PEM fuel cell with metal foam as the flow distributor. *Int. J. Hydrogen Energy* **2012**, *37*, 13060–13066. [CrossRef]

15. Shin, D.K.; Yoo, J.H.; Kang, D.G.; Kim, M.S. Effect of Cell Size in Metal Foam Inserted to the Air Channel of Polymer Electrolyte Membrane Fuel Cell for High Performance. *Renew. Energy* **2017**, *115*, 663–675. [CrossRef]
16. Tabe, Y.; Nasu, T.; Morioka, S.; Chikahisa, T. Performance characteristics and internal phenomena of polymer electrolyte membrane fuel cell with porous flow field. *J. Power Sources* **2013**, *238*, 21–28. [CrossRef]
17. Lefebvre, L.-P.; Banhart, J.; Dunand, D.C. Porous metals and metallic foams: Current status and recent developments. *Adv. Eng. Mater.* **2008**, *10*, 775–787. [CrossRef]
18. Kim, M.; Kim, C.; Sohn, Y. Application of Metal Foam as a Flow Field for PEM Fuel Cell Stack. *Fuel Cells* **2018**, *18*, 123–128. [CrossRef]
19. Fly, A.; Butcher, D.; Meyer, Q.; Whiteley, M.; Spencer, A.; Kim, C.; Shearing, P.R.; Brett, D.J.L.; Chen, R. Characterisation of the diffusion properties of metal foam hybrid flow- fields for fuel cells using optical flow visualisation and X-ray computed tomography. *J. Power Sources* **2018**, *395*, 171–178. [CrossRef]
20. Fly, A.; Meyer, Q.; Whiteley, M.; Iacoviello, F.; Neville, T.; Shearing, P.R.; Brett, D.J.L.; Kim, C.; Chen, R. X-ray tomography and modelling study on the mechanical behaviour and performance of metal foam flow-fields for polymer electrolyte fuel cells. *Int. J. Hydrogen Energy* **2019**, *44*, 7583–7595. [CrossRef]
21. Calvo, S.; Beugre, D.; Crine, M.; Léonard, A.; Marchot, P.; Toye, D. Phase distribution measurements in metallic foam packing using X-ray radiography and micro-tomography. *Chem. Eng. Process. Process Intensif.* **2009**, *48*, 1030–1039. [CrossRef]
22. Wallenstein, M.; Hafen, N.; Heinzmann, H.; Schug, S.; Arlt, W.; Kind, M.; Dietrich, B. Qualitative and quantitative insights into multiphase flow in ceramic sponges using X-ray computed tomography. *Chem. Eng. Sci.* **2015**, *138*, 118–127. [CrossRef]
23. Brackbill, J.U.; Kothe, D.B.; Zemach, C. A continuum method for modeling surface tension. *J. Comput. Phys.* **1992**, *100*, 335–354. [CrossRef]
24. *ANSYS FLUENT 12.0 Theory Guide*; ANSYS Inc.: Canonsburg, PA, USA, 2009.

© 2019 by the authors. Licensee MDPI, Basel, Switzerland. This article is an open access article distributed under the terms and conditions of the Creative Commons Attribution (CC BY) license (http://creativecommons.org/licenses/by/4.0/).

MDPI
St. Alban-Anlage 66
4052 Basel
Switzerland
Tel. +41 61 683 77 34
Fax +41 61 302 89 18
www.mdpi.com

Energies Editorial Office
E-mail: energies@mdpi.com
www.mdpi.com/journal/energies

www.ingramcontent.com/pod-product-compliance
Lightning Source LLC
LaVergne TN
LVHW070553100526
838202LV00012B/458